T0133286

SATURN

Kosmos

A series exploring our expanding knowledge of the cosmos through science and technology and investigating historical, contemporary and future developments as well as providing guidance for all those interested in astronomy.

Series Editor: Peter Morris

Already published:

Jupiter William Sheehan and Thomas Hockey
Mercury William Sheehan
The Moon Bill Leatherbarrow
Saturn William Sheehan
The Sun Leon Golub and Jay M. Pasachoff

Saturn

William Sheehan

REAKTION BOOKS

To Mike Conley

Published by Reaktion Books Ltd
Unit 32, Waterside
44–48 Wharf Road
London N1 7UX, UK
www.reaktionbooks.co.uk

First published 2019
Copyright © William Sheehan 2019

All rights reserved

No part of this publication may be reproduced, stored in a retrieval system, or transmitted, in any form or by any means, electronic, mechanical, photocopying, recording or otherwise, without the prior permission of the publishers

Printed and bound in China by 1010 Printing International Ltd

A catalogue record for this book is available from the British Library

ISBN 978 1 78914 153 5

CONTENTS

Eclipse of the Sun by Saturn, 13 July 2014. The Earth appears as a tiny dot between the two diffuse outer rings, on the right side of Saturn at about four o'clock.

Preface

Saturn is not the largest of the planets, nor the smallest; it is hardly the most Earthlike; not the nearest nor the farthest, not the hottest nor the coldest. It is not even the only one with rings. However, it is the showcase of the solar system: arguably, it is by far the most stunningly beautiful object in the solar system.

The prolific nineteenth-century astronomy writer Richard A. Proctor, whose *Saturn and Its System* was first published in 1865, found it necessary to apologize for devoting an entire book to one planet.

> It might appear, at first sight, that a single planet, however interesting or elaborate the scheme of which it is the centre, should rather be made the subject of a chapter than of a volume, even of the moderate dimensions of the present.[1]

By way of justification, he pointed out that much of what he wrote about Saturn was applicable, with suitable changes of detail, to the other planets of the solar system.

What Proctor wrote now seems rather quaint. Thanks to the efforts of another century and a half of telescopic observations from the Earth, fly-by missions (by *Pioneer* 11 in 1979, and *Voyagers* 1 and 2 in 1980–81) and an orbiter mission (*Cassini*, which circled the planet between 2004 and 2017), we now know far more about

Saturn and its system of rings and satellites than was known, in his time, about the Earth. But it is also true that much of what we have learned is applicable – with suitable changes of detail – to the other planets, most notably Saturn's fellow and sunwards giant, Jupiter, and to the outer giants, Uranus and Neptune, which have rings and icy satellites. Proctor saw in Saturn's ring a vision of the nebular hypothesis of Pierre Simon de Laplace (1749–1827), in which the planets themselves formed from a ring of material surrounding the proto-Sun. In fact, we now know that the rings are debris left over from the break-up of several small moons. But the nebular origin of the solar system has now – with suitable changes of detail – been largely validated, and Saturn (with Jupiter, Uranus and Neptune) was a major player in the earliest acts of the cosmic drama, in which migrations of these gas giants inwards (in the case of Jupiter) or outwards (in the cases of Saturn and the others) created havoc by sending blizzards of material across the solar system, where it pummelled the Earth and Moon during the so-called Hadean era of geological history (occupying the first 700 million years after the birth of the Sun and planets), before they settled into their more or less staid and equable courses.

We now know a great deal about Saturn's atmosphere, as well, including its giant storms and brilliant auroras, and interior (as a gaseous planet, it has no surface, as such). We have begun to explore its fascinating retinue of moons – one of which, Titan, was the first found to have an atmosphere, and another of which, Enceladus, has watery oceans below an icy rind and is seen as one of the most promising places in the solar system beyond Earth to search for life. But it is the magnificent ring system that is Saturn's signature and most sublime feature. Even in a small telescope, it is stunning. With the possible exception of the craters and mountains of the Moon, no other first sight through a telescope has more power to trigger awe and wonder. The rings, appearing so smooth and perfect from afar, as if fashioned on a lathe, are, on closer acquaintance, seen to be a congeries of icy particles, the interactions they undergo among themselves and in response to the tug-of-war between Saturn and its moons making them a veritable laboratory of celestial mechanics. Within the limits of the orbits they describe, they are sculpted into an array of intricate and beautiful forms. Nature ever surprises and delights.

Saturn, overexposed at lower left, passes the diffuse nebulae M8 and M20, 7 August 2018.

ONE

A Pale Yellow Star

Saturn appears to the naked eye as a 'star' of pronounced yellowish hue. It is a giant planet, second only to Jupiter in our solar system in terms of size, and the most remote of those known to the ancients, wandering leisurely among the background stars and completing each orbit round the Sun in 29½ Earth years.

Though no account of its discovery, historical or traditional, exists, as mentioned, it was already known in ancient times. Though less striking than the other planets (except shy and Sun-hugging Mercury), it must have revealed itself early on as a 'wanderer', owing to the fact that it kept to the much-scrutinized region of sky along Earth's orbital plane, or ecliptic, where the Sun, Moon and planets move. In addition, its occasional appearance near a bright star, or close to the Moon, must have centred attention on it, and helped disclose its slow movements and brightness changes over time. Its light is dull, livid and a pale-yellowish colour – the colour of lead (II) oxide (massicot). There is nothing in its naked-eye appearance to suggest that it is one of the most magnificent objects in the heavens. Instead, it has been regarded as an unlucky planet, baleful and malefic.

Prognosticators

Though the discovery of the five naked-eye planets is lost in prehistory, we know, fortunately, quite a lot about the beginnings of the science of astronomy, which took place some 5,000 years ago in the fertile lands between the Tigris and Euphrates rivers in ancient Mesopotamia – in what is now Iraq. In contrast to the great civilization of Egypt, where the annual inundation of the Nile produced predictably and reliably fertile fields, in Mesopotamia conditions for agriculture were more variable. The annual rainfall is low; the ground becomes dry, hard and unsuitable for the cultivation of crops for eight months of the year, while the sluggish flow of water in the two rivers deposits large quantities of silt and elevates the bed to the point where the waters overflow the banks or change their course. Mastery of this challenging situation was achieved only through the creation of an extensive system of artificial canals. The organization of the irrigation networks required coordinated effort on a hitherto unattempted scale and led to the rise of a powerful ruling class and the invention of a script.

In Mesopotamia, these developments seem to have coincided with the settlement, in the southern part of the country, of the Sumerians in about 3000 BCE. The regulation of the calendar became one of the principal functions of the Sumerian priests, who never adopted the solar calendar, as the Egyptians did, and indeed the lunar calendar has remained important ever since in that part of the world. The priests looked out from their seven-level terraced pyramids (ziggurats) for the first appearance of the thin crescent Moon, which marked the beginning of the new month. (The massive ziggurat of the ancient Sumerian city of Ur, built in the period 2112–2095 BCE, is the most famous.) They were also charged with adding, every two or three years, a thirteenth month so that their calendar remained in synch with the seasons and religious festivals.

Later, as the Sumerians merged with the Semitic population of Akkad to form the Sumer-Akkadian Empire, and still later, with the emergence of the Babylonian Empire, the business of the priest-scribes was greatly expanded. In addition to determining the beginning of the new month, as before, and adding intercalary months, they began paying close attention to eclipses, and to phenomena involving the planets – their heliacal risings and settings (that is, first visibility before or after the Sun). They took note of Mercury and Venus' shuttling back and forth between the morning and evening skies, and of the retrograde (backwards) motions of Mars, Jupiter and Saturn around the times they appeared opposite to the Sun in the sky (in *opposition*). The priest-scribes also took note of the planets' changes in brightness; their conjunctions with the Moon, stars and constellations and with each other; and even their appearances within halos about the Moon, and to meteorological phenomena. They were concerned with everything that happened in the sky.

Their interest was not scientific in any modern sense, but astrological. They did not believe the planets were gods, but 'interpreters', and the behaviour of the planets contained 'omens', signals to the kings in which they expressed their pleasure or displeasure. They saw a correspondence between what happened in the heavens and what happened on the Earth (the whole principle of astrology): they believed that the occurrence of phenomena in the sky preceding events of importance – such as the deposition of a king, an uprising, a famine or a war – meant the two must be connected, and that if the same sky phenomenon recurred, the event was certain to follow. Thus they were led to keep careful records on cuneiform tablets of observations of the planets' irregular motions among the stars and the almost endless variety of their phenomena, together with terrestrial events presumed to have foreshadowed them.

Many of the early observations were very imprecise, and there was no clear distinction made yet between astronomical and

Saturn glistens above the great ziggurat at Ur.

meteorological phenomena. Clouds and halos stood on an equal footing with eclipses, and there are many records like this: 'Last night a halo surrounded the Moon; Saturn stood within it near the Moon.'[1] It is unlikely that such observations as these would have directly led ancient observers to discover the regularities that underlie the development of mathematical astronomy. Rather, it is likely that, as historian of astronomy Antonie Pannekoek (1873–1960) supposed, the regularities in the planetary phenomena gradually 'imposed themselves' upon the observers, arousing expectations, which developed into astronomical predictions.[2]

The periodicities thus discovered make up the backbone of Babylonian mathematical astronomy. They include the discovery that Venus takes five complete journeys around the zodiac (the *synodic period*) in almost exactly eight years, so that after eight years it is in the same position relative to the Earth and Sun; and that Mars returns to the same relative position after 79 years, Jupiter after 71 years and Saturn after 59 years. It is probably no coincidence that the oldest Babylonian observations of Saturn, dating from the sixth and seventh centuries BCE, are 59 years apart, and so provide the oldest known material from which the Babylonians could have derived the periodicity of Saturn. (Note that none of these periods is quite precise; in the case of Venus, for instance, the periodicity is actually eight years, minus $2^{2}/_{10}$ days, and for Saturn, 59 days, plus 3 days.)

Cycle and Epicycle

The data of Babylonian astronomy was expressed entirely as these arithmetical regularities, which allowed the prediction of omens. The Babylonians had no interest in developing underlying descriptive models of the actual motions of the bodies in the heavens. That was to be the accomplishment of the Greeks, who began to take possession of Babylonian astronomical data following Alexander the Great's conquest of Babylonia in 331 BCE.

In the hands of the Greek geometers, Babylonian data was interpreted geometrically – that is, geometric constructions were produced that would simulate the apparent paths of the Sun, Moon and planets, as projected onto the apparently flat surface of the sky. They made two assumptions: that the Earth was the centre of the system, and that the planets moved in circular paths. Some ingenious schemes were proposed – notably, that of the fourth-century BCE mathematician Eudoxus of Cnidus, who attempted to represent the retrograde movements of the outer planets (including Saturn) by supposing each one to be moving on a set of interested spheres, rather like a compass in a gimbal. However, though Eudoxus' model was, to a certain degree, successful in representing the qualitative form of the motions, it completely failed to explain the planets' variation in brightness. Saturn, for instance, may appear as bright as magnitude –0.4 (rivalling Sirius, the brightest of the fixed stars) or as faint as +1.5 (a little brighter than a star in the Big Dipper). But why, in an Earth-centred scheme, should a planet's brightness vary at all?

Briefly, in about 250 BCE, the brilliant mathematician Aristarchus of Samos experimented with removing the Earth from the centre of the scheme and replacing it with the Sun – thus introducing the first full-fledged heliocentric system – but the idea won little acceptance. In any case, later geometers preferred to keep the Earth at the centre, and in order to account for the retrograde movements and brightness variations, they claimed that the planets moved on circles whose centres were slightly offset from the Earth (eccentric circles). Geometrically equivalent to this, and much more convenient, was the device of the epicycle: a small circle round which the planet moves while pivoting around a larger circle (known as the deferent) centred upon the Earth. It was already in use in the time of Apollonius of Perga (*c.* 262–*c.* 200 BCE), best remembered today for his work on the conic sections; it is even possible that Apollonius himself was responsible for introducing it. However, it came to its fullest elaboration at the hands of Claudius Ptolemy (*c.* 100–*c.* 170 CE), a Greek who spent his entire career at Alexandria in Egypt and whose book, usually known by the Latin translation of its Arab name, *Almagest*, represents the culmination of Greek – and hence also Babylonian – astronomy.

The Ptolemaic system of epicycles and deferents centred on the Earth has been ridiculed as clumsy and artificial since at least the time of Alfonso X of Castile, nicknamed 'the Wise' (1221–1284), who is reputed to have said, 'If the Lord God almighty had consulted me before embarking upon the Creation, I should have recommended something simpler.'[3] Although the Ptolemaic system is rather complicated, remember that Ptolemy's purpose was not to describe the actual paths of the planets in space but to provide a calculating machine. As Harvard historian of astronomy Owen Gingerich notes,

> Basically, for the first time in history (so far as we know) an astronomer has shown how to convert specific numerical data

> into the parameters of planetary models, and from the models has constructed a . . . set of tables which allow the positions of the planets to be calculated often to within ten minutes of arc, or to well within the limits of accuracy of the measurements possible at the time.[4]

This, in itself, was a gigantic achievement – and, incidentally, a great boon to astrologers, who then, as now, were among the chief users of planetary theory.

We know from another of his books, the *Tetrabiblos*, or 'Four Books', that Ptolemy, like others of his time, took divination by means of the stars quite seriously, and apparently regarded

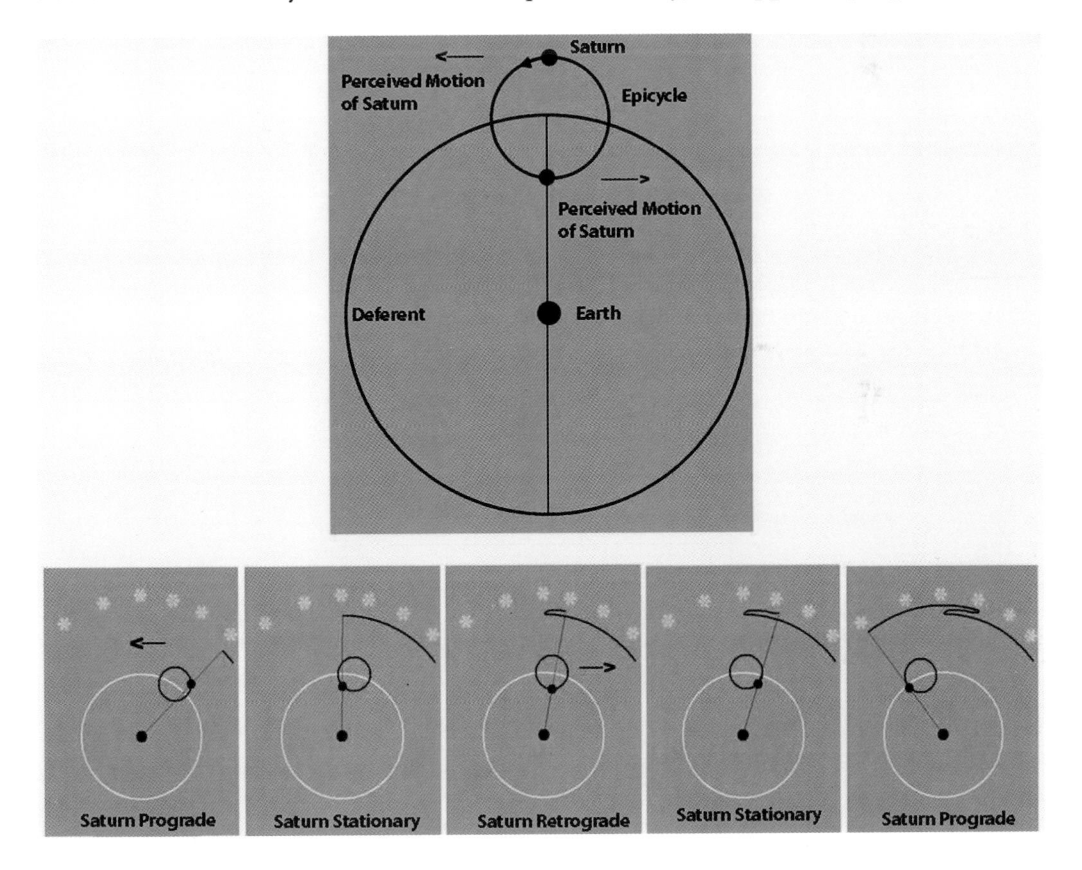

Ptolemaic system. Saturn moves around a small circle, known as the epicycle, which in turn pivots around a larger circle centred on the Earth.

T 536

Francesco Goya, *Saturn*, 1820–23, mixed method on mural transferred to canvas.

astrology as a branch of applied mathematics. His views about epilepsy are typical: 'epilepsy generally attaches to all persons born while Saturn and Mars may be in angles' with the eastern horizon (when Saturn rules the day and Mars rules the night); but if the converse happens (especially if these planets are in Cancer, Virgo or Pisces), 'the persons born will become insane. And they will become demoniac, and afflicted with moisture, of the brain.'[5] With this passage, we remind ourselves that, with Ptolemy, we come to the end of the classical period of learning, and enter upon the dark ages, at least in Europe.

During all this time, Saturn's dull orb, travelling with nearly imperceptible motion, marked the penultimate frontier of the universe, nested between the sphere of Jupiter and that of the fixed stars. It also continued to retain its traditional character as a bad omen. Camille Flammarion (1842–1925) said of Saturn,

> The slowness of its motion and the tint of its light made it for the ancients an unlucky planet. Saturn was, indeed, considered as the gravest and slowest of stars, a god dethroned and banished into a sort of exile.[6]

This recalls an ancient myth: Saturn was the upstart who murdered and castrated his own father, Uranus, and then attempted to safeguard his throne by eating his own children. He was ultimately dethroned, along with the Titans, by Jupiter and the Olympians.

Old, baseless ideas never die; at best, they fade away. Even as late as the nineteenth century, according to Flammarion in his *Popular Astronomy*, the French novelist Victor Hugo (1802–1885) stated, 'in his opinion, Saturn could only be a prison, or a hell'.[7] Perhaps Hugo never looked at Saturn through a telescope: had he done so, he would surely have realized that it is neither prison, nor hell; it is roofed by crystalline rings, a beatific vision.

Farthest Outpost from the Sun

It was a man of the Renaissance, Nicholas Copernicus (1473–1543), canon of Frombork Cathedral, in what is now Poland, who, finding himself irresistibly attracted to the heliocentric system which had been prematurely set forth by Aristarchus, took up the challenge of making it the basis of predictions of the motions of the planets that would equal those based on the Ptolemaic scheme. In order to achieve this, Copernicus fell back on the machinery of epicycles, but he had taken the decisive step: his fully elaborated theory was published in his great book *De Revolutionibus orbium caelestium* (On the Revolutions of the Heavenly Spheres) in 1543 – it is said that the first copies reached him when he was on his deathbed. Saturn, which had hitherto occupied the farthest sphere beneath the fixed stars in the geocentric system, now became the outermost guardian of what, for the first time, became the solar system.

Though acceptance of the book was gradual, Copernicus had fired a shot across the bows, from which there would be hesitation but no turning back. Further progress depended not on theoretical refinements of the epicyclic scheme such as those to which Copernicus had devoted decades of his life, but on better observations, which were to be the legacy of the Danish nobleman Tycho Brahe (1546–1600).

Tycho was the son of Otto, governor of Helsingborg Castle on the other side of the Orsund (Sound) between Denmark and Sweden. Otto arranged with his childless brother Joergen to give Tycho up to Joergen to raise as his own, and frankly, Tycho seems to have been none the worse for the arrangement: Joergen was extremely well-to-do, and doted on his adopted son. He could afford to give him the best education money could buy, and when Tycho was thirteen, he was sent to the University of Copenhagen to begin the study of law. A decisive event in his life took place on

Imaginary view of Saturn's ring, as seen from the surface, from Camille Flammarion's *Les Terres et ciel* (1884).

21 August 1560, when an eclipse of the Sun occurred, just as astronomers had predicted. According to his biographer J.L.E. Dreyer, 'Tycho thought of it as something divine that men could know the motions of the stars so accurately that they could long before foretell their places and relative positions.'[8] He procured for himself a copy of Ptolemy's *Almagest* and worked through it, and acquired a celestial globe and star maps published by Albrecht Dürer, which he used to learn all the constellations. He also began to check the published predictions of planetary positions by lining up a planet and two stars by means of a taut string and estimating the positions of the planets from the positions of the stars on his little globe.

Tycho looking up at the conjunction of Jupiter and Saturn, August 1563.

The next turning point for Tycho occurred in August 1563, and Saturn played a role. From Leipzig, he observed a 'Great Conjunction' of Jupiter and Saturn with a pair of large compasses. In his notebook, he recorded on 18 August,

> The distance between Jupiter and Saturn is a little greater than that between [eta and zeta Auriga] and less than between the two in the front right foot of the Great Bear [iota and kappa Ursa Majoris], but nearer to the interval of those in the Bear . . . The place of Saturn is south of the straight line from Jupiter to Venus, Saturn being further south than Jupiter.[9]

Clearly, he was already striving for meticulous accuracy and soon afterwards showed from his observations that the tables of the astronomers – both those based on Ptolemy and newer ones based on Copernicus – were widely in error. Thus Dreyer wrote that 'at age sixteen Tycho's eyes were opened to the great fact, which seems to us so simple to grasp, but which had escaped the attention of all European astronomers before him, that only through a steadily pursued course of observations would it be possible to obtain a better insight into the motions of the planets.'[10]

Tycho had discovered his life's work, and when, soon afterwards, Joergen died, there was nothing to stand in his way. He observed a 'new star' in Cassiopeia in 1572 at his uncle Steen Bille's estate at Herrevad Abbey (a former Cistercian monastery now in southern Sweden but then in Denmark). He then wrote a book about it, in which he demonstrated that the star was exceedingly remote – far beyond the sphere of the Moon, and even of that of Saturn, and to all intents and purposes located in the sphere of the 'fixed stars'. (The 'new star' was what we now know to have been a supernova – the explosion of a massive star.) The book made Tycho famous, and led the Danish king Frederick II to grant him the island of Hven (a name Tycho always insisted meant the 'island of Venus'), in the sound between Kattegat and the Baltic Sea, where he set up a Baroque 'castle of the heavens', Uraniborg – it looked rather like a gingerbread house with an onion dome and cylindrical towers. Here, Tycho set up a gallery of instruments – sextants and quadrants, each one with open sights, since they were meant to be used to make naked-eye observations.

Tycho's measures of the planets, made relative to the Moon or standard stars which he catalogued, were to be his greatest legacy. For Saturn alone, there are hundreds, made on more than four hundred dates. Here are some examples:

6 January 1587. The Moon came very near Saturn this evening. At $10^{h}\ 4^{m}$ when the eye of the Bull (Aldebaran) was distant from the meridian towards the west . . . then the two horns of the Moon being in quadrature were in exact alignment with Saturn as well as could be seen.

12 February 1592. Saturn's nearest distance from the limb of the gibbous Moon is 9′ of arc or 10′ at most.

29 November 1594. 2½ hours after midnight. Saturn is seen to be stationary and beginning to regress on the straight line leading from Cor Leonis (Regulus) to that which is the unnamed and brighter object below the Great Bear's tail (Cor Caroli). Saturn was distant from Regulus not much more than a degree towards the northeast.[11]

Tycho's grandiose project was nothing less than to determine the structure of the universe. However, although his observations, especially of Mars near its opposition in 1583, seemed to rule out the Ptolemaic system, he did not follow Copernicus all the way; the idea of the 'heavy and sluggish Earth' moving through space was one objection, and there were others. In the end, he devised his own system, in which the planets, excepting the Earth, travelled around the Sun, while the Sun and Moon travelled around a stationary Earth.

Tycho was not as good a landlord as he was an astronomer, and eventually his relations with both his subjects and the Crown deteriorated to the point where he was forced to leave Hven. After a period of wandering, he settled in Bohemia, at first at Benateky Castle, on the River Yser 35 km northeast of Prague, and then in Prague itself, lured by the Holy Roman Emperor Rudolf II. Meanwhile, he made the acquaintance of a young German mathematician, Johannes Kepler (1571–1630), who was, in every way, a different sort of man from Tycho. In contrast to the Danish nobleman, who had every advantage, Kepler grew up in poor, even abusive, circumstances. His father was a ne'er-do-well tavern keeper, who removed him from school and employed him as a potboy, and eventually abandoned the family; no one knows what became of him. His mother was later to be placed on trial as a witch. But genius prevailed. Kepler had a brilliant mind, and although at first he hoped for a career as a Lutheran pastor (he was deeply religious), he was not immediately able to find such a position. Instead, he was offered a post as a mathematics teacher at the University of Graz. The university was under the control of the Catholic Church, but Protestants were not yet banned from Graz – as they were to be soon afterwards – so he accepted.

In the middle of a lecture he was giving in July 1595, Kepler had a flash of insight that would shape the rest of his career. As in the case of Tycho, it involved Saturn – specifically, the planet's conjunctions with Jupiter, which occur at intervals of sixty years. Noting that these close pairings of the planet take place at successive points round the zodiac separated by some 240°, Kepler drew one circle to represent the zodiac, within which he marked off the positions of the successive conjunctions, and connected these points with line segments so as to produce a series of triangles. Their overlapping figures defined two circles, of which the inner circle seemed to stand in relation to the outer in nearly the same proportion as the orbit of Jupiter to that of Saturn.

Seeing this, he took the next step and placed a square inside the orbit of Jupiter, hoping that the inscribed circle would have the dimensions of the orbit of Mars. He did not obtain the anticipated result. Undaunted, he next tried fitting the series of polyhedra known as the 'Platonic solids' (since they are mentioned in Plato's *Timaeus*) inside a series of interested spheres representing the orbits of the planets. The five Platonic solids – and there are neither more nor fewer – are the cube, octahedron, icosahedron, tetrahedron and

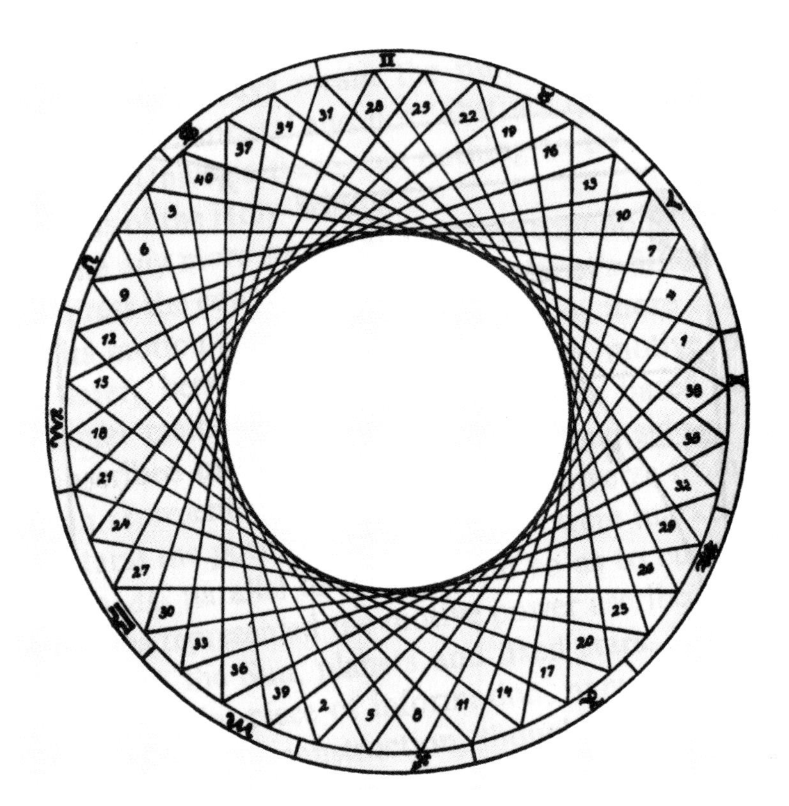

Kepler's figure from *Mysterium cosmographicum*, showing successive positions of the conjunctions of Jupiter and Saturn.

dodecahedron. From their properties as discussed in Euclid, Kepler was able to calculate the dimensions of the orbits. Though the results were far from perfect, they were at least of the right order. Admittedly, the whole scheme seems crackbrained by today's standards, but Kepler was only 24 years old, and thought that he now understood the secret of the structure of the universe as it had existed in the mind of God, including why there were five, and only five, planets. He wrote a book on the subject, *Mysterium cosmographicum*, in 1596, in which he presented his theory, and sent a copy to Tycho. Tycho – though not entirely convinced – recognized the ingenuity that had gone into it, and since he was looking for an assistant at the time, invited Kepler to join him in Bohemia. Kepler accepted.

Neither man was easy to get along with, and their relations were strained from the first. However, Tycho had less than a year to live: on 24 October 1601 he died, apparently from a bladder obstruction. On his final delirious night, he repeated over and over again, 'Let me not have lived in vain.' Although Tycho hoped that his observations would be used to establish the correctness of his – the Tychonic – system, Kepler had been a confirmed Copernican since 1597, and eventually, after a series of backbreaking calculations, established the result known as Kepler's first law of planetary motion: that the actual shape of the orbits of the planets is elliptical, with the Sun at one focus and the other empty.[12]

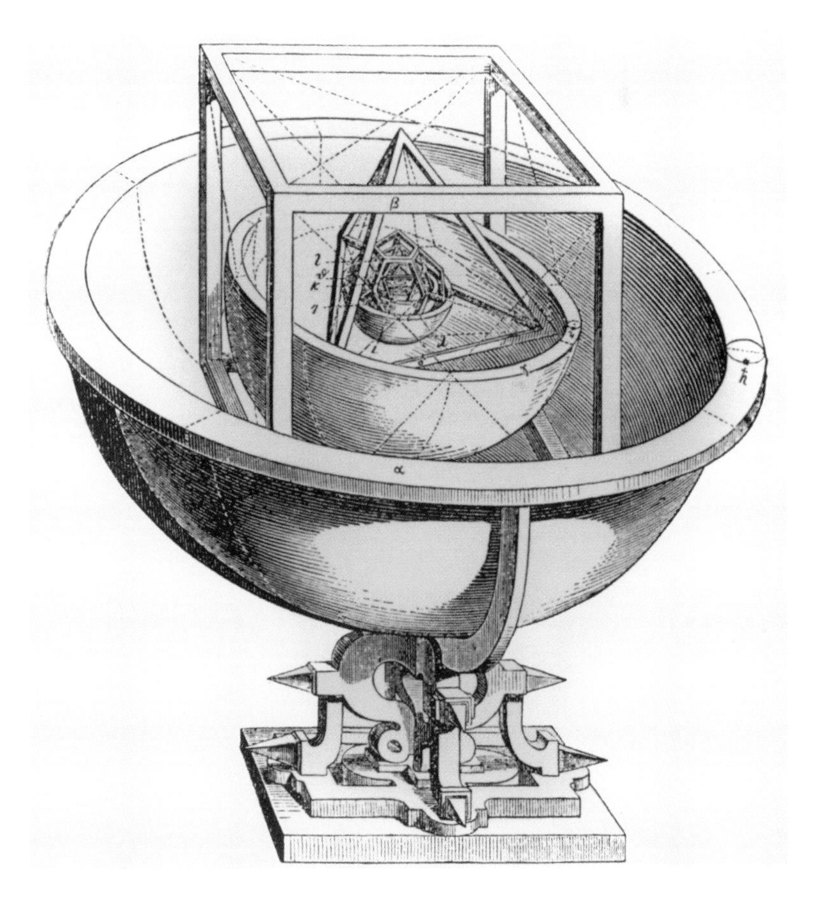

The solar system, as envisaged in the *Mysterium cosmographicum* (1596).

This was a result, published in 1609, that could never have been anticipated by the ancient astronomers. It was due, first, to the meticulous accuracy of Tycho's observations, and second, to the fact that when he joined Tycho, Kepler was, by sheer good luck, set to analysing the motions of Mars, whose orbit is more eccentric (varies more from the circular) than any other except Mercury's. (Its eccentricity is 0.093, compared to 0.007 for Venus, 0.017 for the Earth, 0.049 for Jupiter and 0.056 for Saturn; note that the eccentricity of a circle is defined as 0.00.) Had he started with another planet – say Venus – he would never have made this momentous discovery.

Based on Kepler's work, Isaac Newton would later establish the inverse square law of gravitation, which explains, in detail, the motions of the planets and their moons, of which, arguably, none is more intricate or beautiful than those involving Saturn.

We have here been able to give only the briefest account of the planetary motions. An account such as this can, of course, give little indication of the difficulties faced by mathematical astronomers. Each planet pulls upon the next, producing perturbations of its movements; the satellites – and even the largest asteroids – must be taken account of. In particular, the effects of Jupiter and Saturn, by far the most massive planets of our solar system, dominate the scene. We will have more to say about all this later, but for the present, let us turn our attention from the movements of Saturn, to consider it as a world.

TWO

A Strange Ringed World

The ancients could never have anticipated the elliptic orbits of the planets, but even less conceivable to them would have been what Saturn looks like through a telescope. By a striking coincidence, in the same year that Kepler published his discovery of the elliptic shapes of the planetary orbits, Galileo Galilei (1564–1642), a professor of mathematics at the University of Padua, began to observe the heavens with an instrument he called a perspicillium, and which Kepler subsequently renamed the telescope. Among the many revelations made by Galileo and his successors thanks to the telescope, none were more remarkable than those regarding Saturn.

Though paltry by today's standards, Galileo's telescopes were the best of their time, and by November 1609 he had produced one magnifying 20×, with which he made the well-known discoveries of mountains and craters in the Moon and the four large satellites of Jupiter.

By the following summer, Galileo had a more powerful telescope, magnifying 32×, and pointed it at Saturn. He described what he saw in a letter, dated 30 July 1610, to Belisario Vinta (1542–1613), Counsellor and Secretary of State to the Grand Duke of Tuscany:

> I have discovered a most extraordinary marvel . . . The fact is that the planet Saturn is not one alone, but is composed of three,

> which almost touch one another and never move nor change with respect to one another.[1]

Galileo assumed, naturally enough, that the two lesser bodies were satellites, and announced to Giuliano de' Medici (1574–1636), the Tuscan ambassador to Prague, 'So! we have found . . . two servants for this old man, who help him to walk and never leave his side.'[2]

If, however, they were satellites, they were strange satellites indeed: two and a half years later – at the end of 1612 – Galileo examined Saturn again, and found the satellites had vanished without a trace. Saturn now presented a gold-tinted orb as round and complete as that of Jupiter. 'What can be said of this strange metamorphosis,' he wrote to the German banker and amateur astronomer Mark Welser (1558–1614) on 1 December 1612,

> that the two lesser stars have been consumed, in the manner of the sunspots? Has Saturn devoured his children? Or was it indeed an illusion and a fraud with which the lenses of my telescope deceived me for so long – and not only me, but many others who have observed it with me?[3]

Two of Galileo's drawings from 1610.

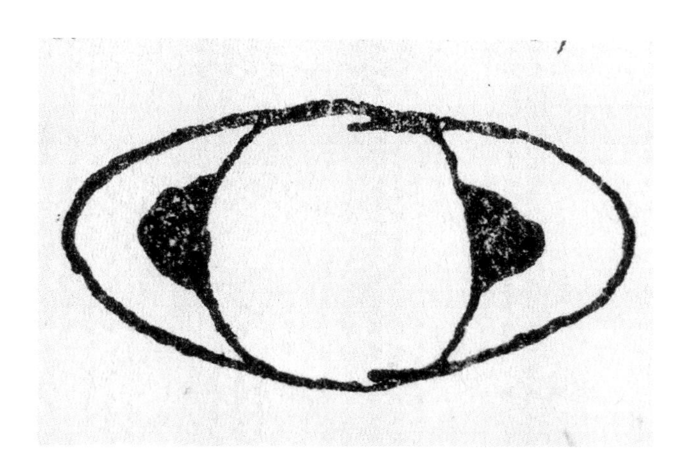

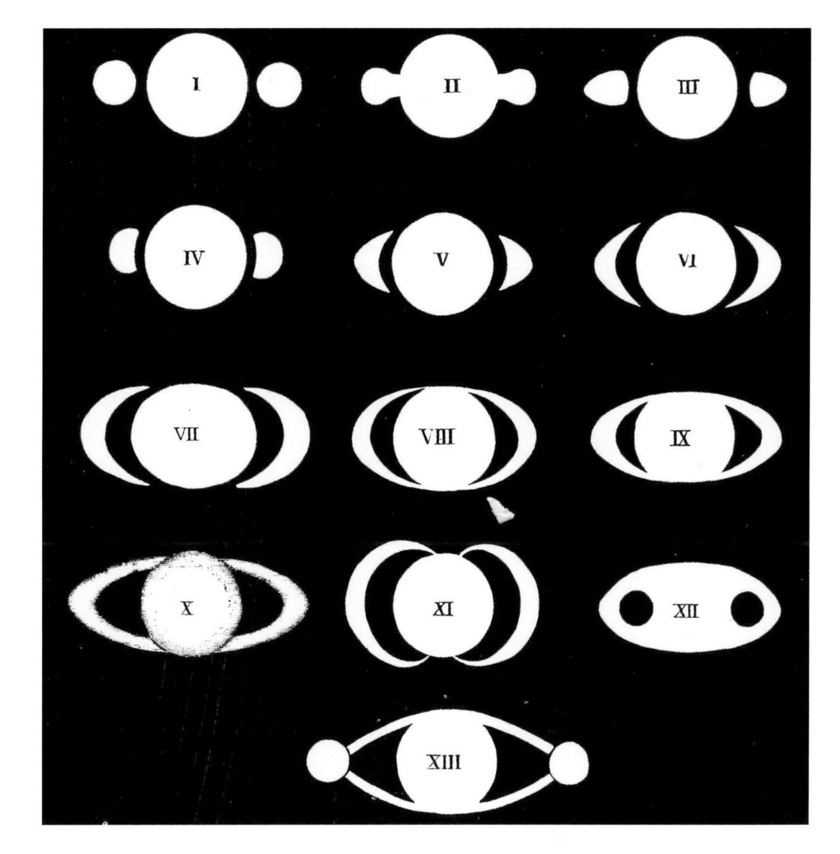

Early drawings of Saturn: I Galileo (1610); II C. Scheiner (1614); III G. Riccioli (1641 or 1643); IV–VII J. Hevelius (theoretical forms); VIII, IX Riccoli (1648–50); X E. Divini (1646–8); XI F. Fontana (1636); XII G. Biancani (1616), P. Gassendi (1638–9); XIII Fontana and others at Rome (1644–5).

Galileo's perplexity was extreme when later the satellites reappeared. In his drawing of 1616, the tri-form appearance of 1610 has been transformed into a globe with looped handles (*ansae*). One suspects that if Galileo had only seen the planet thus from the first, he might well have realized the truth at once.

As it was, the riddle of Saturn would continue to challenge astronomers for another half-century. Among the ideas put forward were that the equator of Saturn was a torrid zone giving off vapours, or that there were two dark satellites periodically passing in front of or behind two lighter ones. And these were the most plausible! The solution of this astronomical

Rubik's cube awaited another man of mathematical and scientific genius, Christiaan Huygens (1629–1695) of the Netherlands.

Christiaan was the second son of Constantijn Huygens, Secretary of State to three successive princes of Orange, and a gifted linguist, musician and poet. He grew up a sickly and withdrawn youth, addicted to study, and intellectually precocious, in the luxurious surroundings of his father's house at Het Plein, the town square of The Hague. He read Galileo's last and greatest book, *Two New Sciences*, which laid the foundations of mechanics, and absorbed the analytical geometry of René Descartes (1596–1650). For the rest of his life, he would insist on mechanical explanations in science.

His father sent him to Leiden University to study law in the hopes that he would follow in his footsteps and pursue a career in statecraft; but like Tycho at the same age, he could not resist his inclination towards other studies, and though he finished his law degree, he did so half-heartedly, and as soon as he could he returned to Het Plein and devoted himself to mathematics, the study of motion and – in the 1650s, in collaboration with his older brother, another Constantijn – to improving the telescope. By March 1655, the brothers had finished a telescope with a lens of ordinary grey-green plate glass 5.7 cm in diameter, and providing a useful magnification of only 50×. It was about as powerful as the department store telescope with which many (including the author) have begun their careers as observers. But though modest by today's standards, it was far superior to the telescopes Galileo had used, and on pointing it through an attic window, Huygens discovered Saturn's large satellite – which the nineteenth-century English astronomer Sir John Herschel (1792–1871) would later name Titan.

Already, Huygens suspected that the strange appearances seen from Galileo onwards were due to the existence of a 'thin, flat ring, nowhere touching'. Huygens published this theory in his *Systema*

Caspar Netscher, *Christiaan Huygens*, 1671, oil painting.

Saturnium, in 1659. A strong believer in the notion that a picture is worth a thousand words, Huygens presented his theory with an extraordinary diagram, of which graphics design author Edward R. Tufte writes,

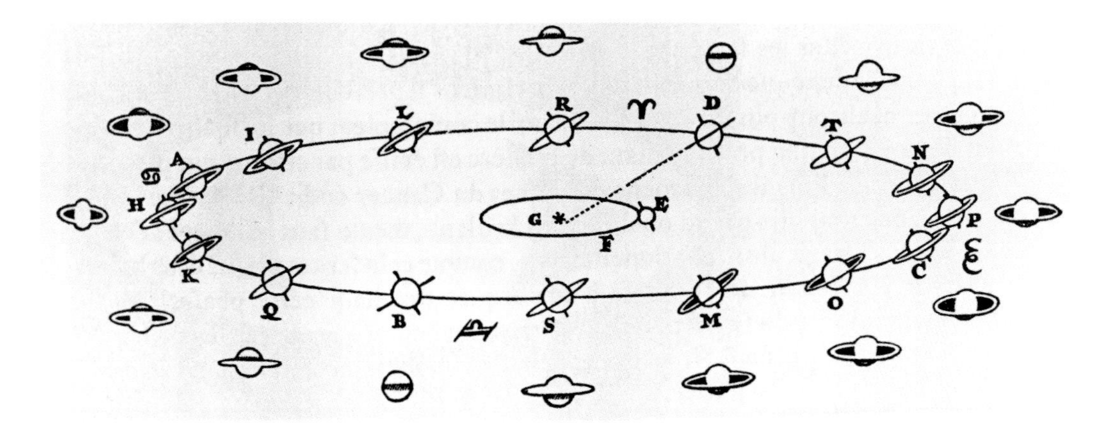
R
D
L
I
T
N
A
H
G
E
F
P
K
C
Q
B
S
M
O

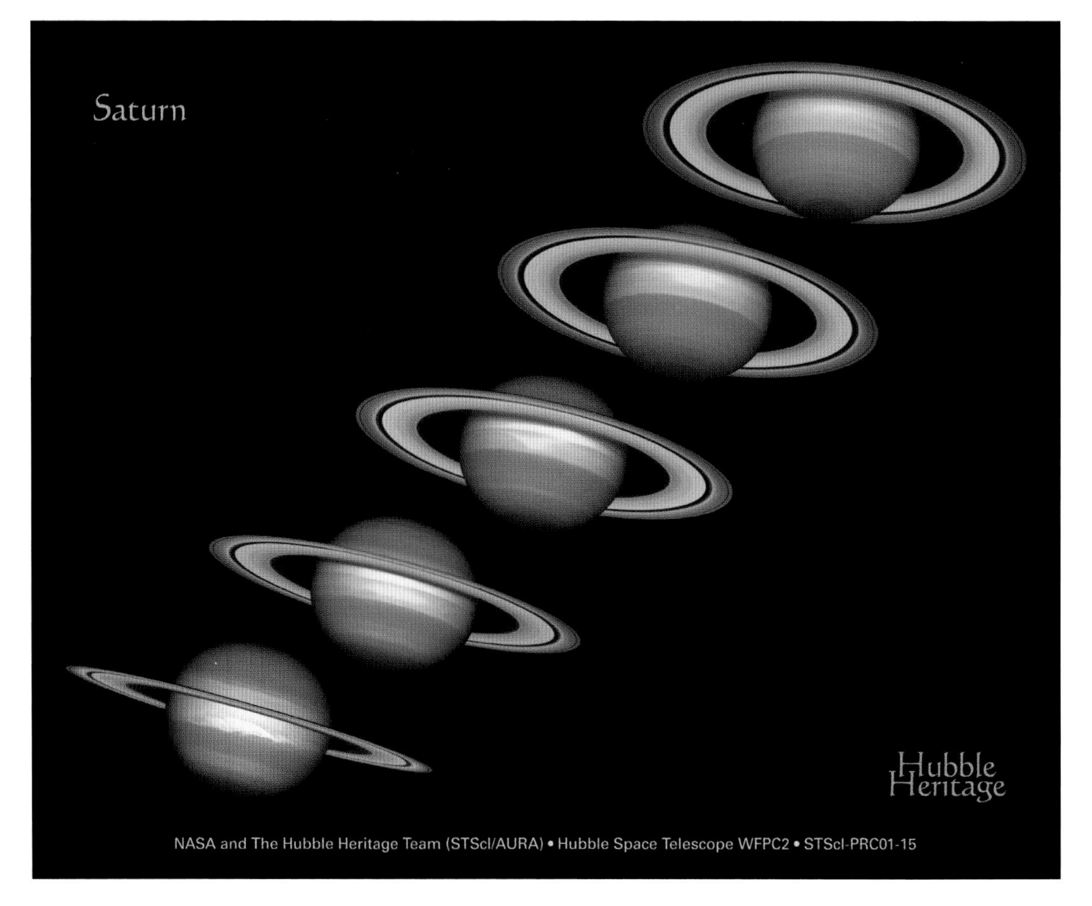
Saturn
Hubble Heritage
NASA and The Hubble Heritage Team (STScI/AURA) • Hubble Space Telescope WFPC2 • STScI-PRC01-15

The inner ellipse traces Earth's yearly journey around the Sun; the larger ellipse shows Saturn's orbit, viewed from the heavens. The outermost images depict Saturn as seen through telescopes on Earth. All told, we have 32 Saturns, at different locations in three-space and from the perspective of two different observers – a superior *small multiple* design.[4]

The plane of the ring is, of course, fixed with reference to the stars, but from the viewpoints of the Sun and Earth, the tilt appears to be continually changing. At alternate intervals of 13 years 9 months and 15 years 9 months, the Sun and Earth cross the plane of Saturn's rings. Because Saturn's orbit is inclined by some 2°.5 to the plane of the Earth's orbit (the ecliptic), the Sun and Earth crossings do not quite coincide. Instead, the Earth crossings – when Saturn's thin rings appear exactly edgewise to us – occur several months before or after the Sun crossings. On these occasions, the Earth passes either once or three times through the ring plane. (Rarely, there is a near-miss, and no ring-plane passage occurs at all.) As we now know, the ring had been edgewise in December 1612, hence its invisibility in Galileo's small telescope.

It may be useful here to list, for reference, the following data:

Inclination of	
Earth's equator to the ecliptic	23°.4
Earth's equator to Saturn's ring plane	6°.6
Saturn's equator to Saturn's orbit plane	26°.7
Saturn's equator to the ecliptic	28°.1
Saturn's orbit plane to the ecliptic	2°.5

Above: Huygens's ingenious and much-praised diagram from his book *Systema Saturnium* (1659). *Below*: images from the Hubble Space Telescope showing aspects of Saturn between 1996 (bottom) and 2000 (top).

Saturn's polar axis is tilted 26°.7 to the plane of its orbit, which is very similar to the Earth's 23°.4. It follows then that Saturn, like the Earth, has marked seasons. For Saturn, the

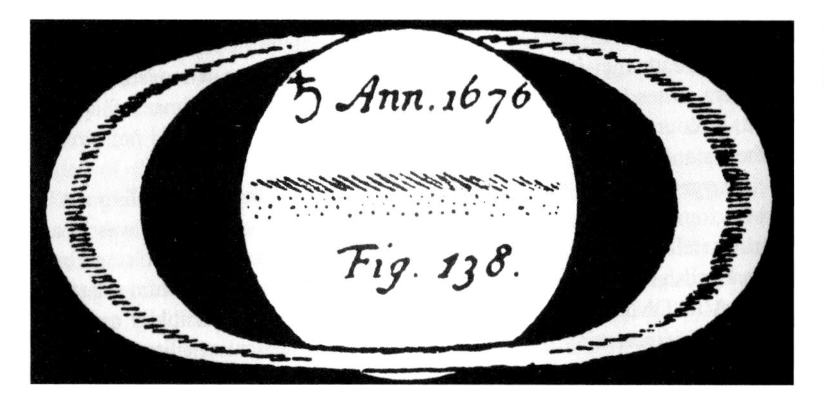

Cassini's sketch of Saturn, 1676, showing the Cassini Division in the ring.

equinoxes correspond to the edgewise ring positions, the solstices to the widest open phases of the rings. However, since the southern hemisphere's summer solstice occurs shortly before perihelion and the northern hemisphere's shortly before aphelion, the southern hemisphere will have shorter but warmer summers than the northern hemisphere. On the other hand, the southern hemisphere's winters will be longer and colder. Since the Sun's distance from Saturn is 10.7 per cent less at perihelion than at aphelion, the difference in solar insolation over the two hemispheres is not insignificant at 20 per cent. We shall consider the implications of this difference later.

Though Huygens would make many more important discoveries in astronomy, mathematics and physics, his discovery of Saturn's ring remains the most famous. His poet father recognized the achievement of his thirty-year-old son, and commemorated it in a couplet:

> *Gloria sideribus quam convenit esse coævam*
> *Et tantum cœlo commoriente mori.*[5]
> (Glory coeval with the skies on high:
> To die, only when heav'n itself shall die.)

The Ring Becomes Plural

What Huygens saw as one ring later observers showed to have a more complex structure. In 1675 Giovanni Domenico Cassini (1625–1712) at the Paris Observatory discovered that Huygens's ring was divided into two by a jet-black line, presumably a true gap, only a half second of arc in breadth (the apparent breadth of a human hair seen at a distance of 25 m). This 4,700-km wide feature is now known as the Cassini Division. (It is a reminder of just how remote Saturn is, and how vast its globe and system of rings, that this hair's breadth of a feature seen in Cassini's telescope – the most powerful of its time – is actually greater in width than the distance between New York and Los Angeles.) Parenthetically, the ring outside the Cassini Division is known as the A ring, with a width of 14,835 km; the ring inside the Cassini Division is called the B ring, and is the widest of the classical rings with a width of 25,554 km.

Henri Testelin (after Charles Le Brun), *Colbert Presenting the Members of the Royal Academy of Sciences to Louis XIV in 1667*, 17th century. Both Cassini and Huygens were members: Cassini stands eighth to the left of Colbert, Huygens just to his right.

Cassini discovered the Division using a telescope made by the Roman instrument-maker Giuseppe Campani (1635–1735). It had a single-element lens 6.4 cm in diameter with a focal length of 6 m, magnified 90× and was the most powerful research instrument of its time. This was, indeed, the era of long telescopes. Though they must have been very difficult to use for protracted observations, their optical performance was better than is often assumed, and would be difficult to surpass even with modern telescopes of similar aperture.

Though as a theoretician Cassini was not in Huygens's class – he never accepted the Copernican theory, instead continuing to espouse that of Tycho Brahe – there is no questioning his skill as an observer, or his great aptitude for making striking discoveries. He discovered four more moons of Saturn, which he called the 'Louisan Stars' (in honour of his patron King Louis XIV; the names by which they are known today were given to them by Sir John Herschel). Iapetus was discovered in 1671, and Rhea the following year. Cassini found Iapetus to be remarkable for the fact that it appears two magnitudes brighter at its western elongation than at its eastern one. He also discovered Tethys and Dione in 1684, with 'aerial' telescopes with focal lengths of 30 m and 41 m (Huygens, by the way, would never see them.)

Cassini and his son Jacques (1677–1756), also an astronomer and a member of a Cassini dynasty at the Paris Observatory that would last until the French Revolution, suggested that the rings were probably formed of an infinity of small planets.[6] That view, however, did not prevail at the time. Instead, Cassini's own nephew Jacques Maraldi (1665–1729), an assistant at the Paris Observatory, was convinced from the telescopic appearance that the ring was not only solid but rigidly so. Even the great telescope-maker and amateur astronomer William Herschel (1738–1822) subscribed for a long time to the solid-ring theory, and refused to admit that even the Cassini Division

Giovanni Domenico (or Jean-Dominque) Cassini, the leading astronomer at the Paris Observatory from 1669. Portrait by Léopold Durange after an old engraving. Note the aerial telescope mounted on the rooftop of the observatory.

was a true void until he had tested it observationally. (He eventually would do so.) There remained, however, a nagging question: how could a solid ring of such dimensions hold together without being ripped apart by centrifugal force?

This question was investigated by the mathematical astronomer Pierre-Simon de Laplace, who, in a celebrated memoir published in 1787, concluded that a single, solid ring as large as Saturn's was an impossibility – quite simply, it could not hold together.[7] If Saturn were assumed to be perfectly round, a solid, uniform ring could indeed continue rotating round it forever; this is, however, a mere mathematical idealization, and in reality the globe of the planet is flattened by rotation into an oblate spheroid, so that any ring, being disturbed by the gravitational attractions of the satellites, other planets and the Sun, would quickly collapse. In the end, Laplace was led to propose a rather unwieldy system of narrow, eccentric and irregular ringlets. To bolster his claim, he recalled various observations, including one by a well-known English instrument-maker, James Short (1710–1768), who once reported seeing many fine divisions in the rings.

According to Saturn historian A.F.O'D. Alexander (1896–1971),

> Laplace's theory, based on mathematical and physical considerations, postulates a ring system which seems much too artificial to have a real and continuing existence, and one

> cannot help being surprised that it should have held the field for half a century. From such a theory to the total abandonment of the idea of solid rings seems but a short step to take.[8]

Whether it was on account of a blind spot, failure of nerve or hesitation in taking a different view from an astronomer as prestigious as Laplace – whatever the reason – the short step was not taken. But there is no doubt that Laplace's idea of many ringlets stimulated observers to scrutinize the rings more closely, and consistent with expectation, a host of minor details began to be reported.

William Herschel – who, incidentally, with the discovery of Uranus in March 1781, had dethroned Saturn from the position it had held since antiquity as the most remote of the planets – made many observations of Saturn with reflecting telescopes having focal lengths of 2.1, 3.0 and 6.1 m. He finally convinced himself that the Cassini Division was indeed a true void in the rings, but could never bring himself to accept Laplace's 'narrow slips of rings'. In a long summary of his Saturn observations published in 1792, he wrote,

> The mind seems to revolt, even at first sight, against an idea of the chaotic state in which so large a mass as the ring of Saturn must needs be, if phenomena like these can be admitted. Nor ought we to indulge a suspicion of this being a reality, unless repeated and well-confirmed observations had proved, beyond a doubt, that this ring was actually in so fluctuating a condition.[9]

Herschel himself had only once seen anything resembling a new division in Saturn's rings, a black 'list', or linear marking, that appeared on the inner edge of Ring B in June 1780. But later observers reported greater success.

Saturn's rings resolved into concentric ringlets, as envisaged by Laplace.

In 1837 Johann Franz Encke (1791–1865), Director of the Berlin Observatory, using the 24-cm refractor later used for the discovery of Neptune, reported a broad, dusky feature in the middle of the A ring. It was definite enough to allow measurement of its position with a filar micrometer, and was soon confirmed by several other observers. Yet questions remained: it was not clear whether it was an actual gap like the Cassini Division or merely a thinning in the ring, or even whether it was a permanent or transient feature. Further study was needed.

As mysterious as they were beautiful, the rings seemed almost to defy the laws of physics: how could solid rings of such dimensions ever remain intact? Summing up his wonder and perplexity on viewing Saturn in a telescope, the American astronomer Ormsby Macknight Mitchel (1810–1862) wrote in 1842,

> The telescope was directed to a small dim star, not far distant from Jupiter, presenting nothing remarkable in magnitude or

The 38-cm Merz and Mahler refractor of the Harvard College Observatory, used to make a series of important observations of Saturn in the late 1840s and '50s.

brilliancy to the unassisted eye. But with a power equal to a thousand eyes, how great the change! An orb of surpassing beauty, encircled by two broad flat rings, and engirdled by no less than seven moons, comes up from out the deep distance to greet the astonished beholder. No person ever beheld this wonderful system for the first time without a burst of admiration. When we connect with the personal inspection of the Saturnian system the facts with reference to its mysterious arrangements, the stability of these two immense flat rings, some two hundred thousand miles [320,000 km] in diameter, upon the exterior, separated from each other, and from the body of the planet, each revolving about the same axis on which the planet rolls, and with a velocity a thousand fold greater than the speed with which the parts of the earth's equator are carried by its diurnal rotation; when we imagine the diversified scenery which is presented by these rings, and by the moons, some rising, some setting, others waxing or waning, some going into or coming out from an eclipse, their vast proportions, the rapidity of their motions and changes, – the mind is overwhelmed in wonder and astonishment.[10]

The Mysterious C Ring

Though Laplace's theory, now over sixty years old, had seemed at least broadly consistent with the observations of Saturn's ring structure, serious doubts began to emerge in 1850. That year, a dusky inner ring was discovered by George Phillips Bond (1825–1865) of the Harvard College Observatory. Using the 38-cm Merz and Mahler refractor – the largest telescope in America at the time, and the equal of the great Pulkova refractor in Russia – Bond had first suspected a penumbral light on the inner edge of Ring B as early as 10 October 1850. This was seen more definitely on 11 November, although the idea that the dusky shading might be a ring was first suggested by a volunteer assistant, Charles W. Tuttle (1829–1881). Before news had crossed the Atlantic, the ring had been seen independently by William Rutter Dawes (1799–1868) with a 16-cm Merz refractor at Wateringbury, near Maidstone, England.

In Dawes's account,

> On the 25th of November, I detected for the first time within the ansa of the ring at both ends while examining the planet with my Munich refractor of [16 cm] aperture. While I was endeavouring to make out what it could possibly be I was interrupted by some visitors . . . The next fine night, the 29th November, I attacked it vigorously, and made it all out, though scarcely able to believe my eyes or my telescope . . . On 2 December Mr Lassell came to see me . . . and the next night, the 3rd, being fine, I prepared to show him this novelty, which I had told him of and explained by my picture; but, naturally enough, he was quite indisposed to believe it could be anything he had not seen in his far more powerful telescope. However, being thus prepared to look for it, and the observatory being darkened to give every advantage on such an object, he was able to make it all out in a few minutes.[11]

The Rev. William Rutter Dawes, known as the 'eagle-eyed', who independently discovered the crêpe ring of Saturn.

Dawes's friend William Lassell (1799–1880), a wealthy brewer and amateur astronomer in Liverpool,

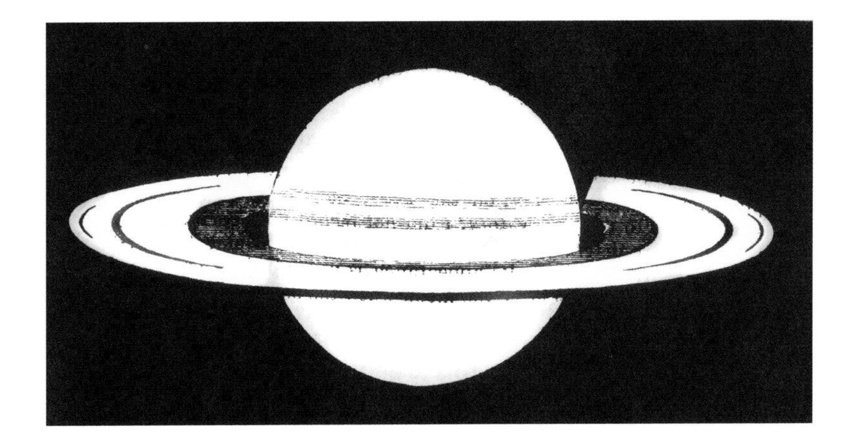

Drawing of Saturn by Dawes, December 1850, showing the crêpe ring and also the Encke division.

on peering through Dawes's telescope first described the dusky ring as looking like a crêpe veil across the planet, and it is still sometimes referred to as the crêpe ring. Its more prosaic and official name is Ring C.

The new ring proved to be relatively easy to see – Dawes, after all, had used a telescope of only 16 cm aperture. This was surprising, given its late date of discovery. How then could it have been missed by such diligent observers as Herschel? The Russian astronomer Otto Wilhelm von Struve (1819–1905) suspected that the ring might have brightened over the years and, on comparing his own measures of the dimensions of the rings with those made by earlier observers such as Huygens and Cassini, further suggested that the B ring was careening inward towards the planet at an alarming rate. According to his calculations, it would come into actual contact with the surface in about the year 2150. He was obviously wrong: there is no possibility of the dimensions of the main rings having changed appreciably since Huygens or Cassini's time, nor has the C ring changed. Indeed, the projection of the dusky inner ring against the ball of Saturn can be traced to observations by Campani in 1664, Robert Hooke (1635–1703) in 1666, Jean Picard (1620–1682) in 1673 and John Hadley (1682–1744) in 1720. (All had assumed it was simply a dusky belt on the globe of Saturn.) When Herschel,

Saturn, with the crêpe ring, as observed by William Lassell with his 61-cm equatorially mounted reflector at Valletta, Malta, in November 1852.

in 1793, sketched the planet, he showed what are easily recognized as three atmospheric belts and a portion of Ring C silhouetted in front of Saturn's globe. The three belts on the upper part of the globe differ in curvature from the fourth, which Herschel faithfully depicts as perfectly concentric with the inner edge of the bright rings' ellipse. Despite this powerful visual cue, Herschel failed to follow the outline off the globe and to see its extension into the space beyond, and so missed an important discovery.

Almost immediately after it was recognized by Bond and Dawes, the crêpe ring was shown to be semi-transparent – the outline of the ball could be seen through it. The result was not easy to reconcile with the idea of solid ringlets. Could it be, then, that the rings were not solid at all, but fluid?

The fluid-ring hypothesis was originally proposed by G. P. Bond himself, almost immediately after the new ring was discovered, in a

William Herschel's drawing of 11 November 1793, showing the feature he called the 'quintuple belt', which includes the three dark belts and two light belts between them. The shadow of the C ring appears as a fourth dark belt in projection against the planet's ball.

paper titled 'On the Rings of Saturn', read before the American Academy of Arts and Sciences in Boston on 15 April 1851.[12] Bond called attention to the fact that observers had tended to disagree as to the number and position of the minor ring divisions, while at times, even with the most powerful telescopes in the world and under perfect conditions of definition, none whatever had been visible. (That, incidentally, would continue to be the case far into the twentieth century.) If the rings consisted of irregular, solid ringlets such as Laplace had proposed, changes were perhaps to be expected; but, Bond implied, they were likely to be violent, since such ringlets would 'become the source of mutual disturbances, which must end in their destruction, by causing them to fall upon each other'. On the other hand, 'the hypothesis that the whole ring is in a fluid state, or at least does not cohere strongly, presents fewer difficulties.' Bond's Harvard colleague the mathematician Benjamin Peirce (1809–1880), in another paper read to the American Academy, attempted to show theoretically that the ring must consist 'of a stream, or of streams of a fluid rather denser than water, flowing around the primary'.[13]

It is well known that expectation can often lead to perception, and this seems to have been the case following the publication of the fluid-ring hypothesis. Using the same telescope with which Bond had discovered the crêpe ring, Tuttle and another volunteer observer, Sidney Coolidge (1833–1863), reported three or four curvilinear markings, which they described as like 'fine divisions or waves'.[14] On 20 October 1851, Tuttle, finding the definition 'admirable', with the planet 'beautifully steady and distinct', was struck by an appearance so remarkable that he called two other observers to the telescope for confirmation. Tuttle wrote, 'There is no question but that B is minutely subdivided into a great number of narrow rings.' Later, Tuttle elaborated on the observation from memory:

> The divisions were not unlike a series of waves; the depressions corresponding to the spaces between the rings, while the summits represented the narrow bright rings themselves. The rings and the spaces between were of equal breadth.[15]

The following year, Lassell, deploying a 61-cm equatorially mounted reflector in the clear air at Valletta, on the island of Malta, saw a series of concentric and deepening bands of shade in the interior of Ring B, which he compared to the steps of an amphitheatre.[16] Dawes, now using a 19-cm refractor by Alvan Clark, also saw the B ring

> decidedly in stripes . . . about one fifth of its breadth from the outer edge very bright; then a lightly shaded narrow stripe;

Charles W. Tuttle's drawing, showing a series of wave-like ripples or fine divisions in the inner part of Ring B.

The firm of Alvan Clark & Sons, established at Cambridgeport, Massachusetts, not far from the Harvard College Observatory, would on five occasions build the world's largest refractor. Its founder, Alvan Clark, had been a professional portrait painter before he became an optician, and produced this exquisite drawing of Saturn with the 38-cm Merz and Mahler refractor at Harvard on 20 December 1853.

> then a lighter stripe; next a considerably darker stripe, then a much darker one extending nearly to the inner edge.[17]

The impression of evanescent structures, like ripples or waves, seemed to be gaining ground.

Like the Siege of Sebastopol

The fluid-ring hypothesis created scarcely a ripple before it was disproved, in favour of what was hailed at once as the definitive explanation of the rings' composition.

Already, two years before the discovery of the crêpe ring, an important insight into the nature of the rings had appeared in an obscure journal, the *Mémoires* of the Montpellier Academy in France. There, Édouard Roche (1820–1883), a mathematician, had shown that a satellite, either fluid or solid, would be disrupted by tidal forces if it approached too close to its primary planet. Assuming an equal density for planet and satellite, he calculated that the critical distance (Roche's limit) was 2.44 times the radius of the planet. For Saturn, this was just outside the outer edge of Saturn's A ring. Roche had foreshadowed modern theories of the rings' genesis, but unfortunately his work remained almost unknown at the time.

Then, in 1855, the University of Cambridge offered as the subject of its 1856 Adams Prize essay competition (named after the university's mathematical astronomer John Couch Adams, famed for the calculations that had led to the discovery of Neptune) the question of whether the rings of Saturn were 'fluid or aeriform, or be supposed to consist of numerous small and unconnected masses'.[18] There was but one entrant, though that one was extraordinary: James Clerk Maxwell, who, when the subject of the prize was announced, was a 25-year-old Fellow of Trinity College, Cambridge.

Maxwell had been born in Edinburgh into a family of comfortable means: his father was of the Clerk family of Penicuik, Midlothian, and his uncle was the 6th Baronet of Penicuik. At an early age, his family moved to Glenlair House, which his parents had built on the Middlebie country estate near Castle Douglas, Kirkcudbrightshire, in southeast Scotland. From a very early age, he demonstrated unquenchable curiosity, and from the age of three, everything that moved, shone or made a noise drew his question, 'What's the go o'that?' (In other words, 'how does that work, exactly?') If not satisfied with the answer, he would follow up with, 'What's the *particular* go of it?'[19] He also had a precocious

James Clerk Maxwell, as a student at Trinity College, Cambridge, at about the time he succeeded in solving the problem of the composition of Saturn's rings.

flair for mathematics: a first paper, on a method for drawing ovals, was published when he was only fifteen. Though he went on to become one of the greatest physicists who ever lived (for his theory of electromagnetism), at the time he took up the Saturn challenge he was not yet well known, and he saw in this an excellent opportunity to build his reputation. Because of his father's ill health,

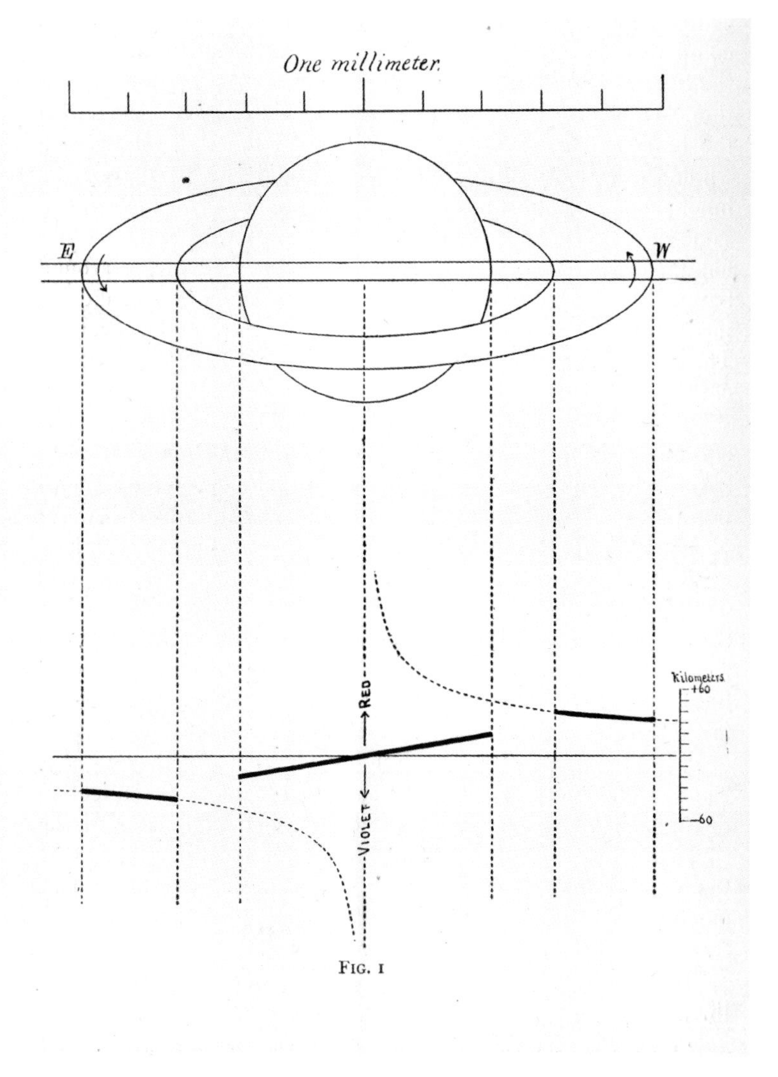

James Keeler's demonstration of the 'meteoric constitution' of Saturn's rings.

he wanted to leave Cambridge and return to his native Scotland, and did so, taking up a position as professor of natural philosophy at Marischal College, Aberdeen, in 1856. Only then did he start work on Saturn, and then only while carrying out other duties, such as teaching and tutoring students. He first showed mathematically that neither solid nor liquid rings would hold together, and then proved that the rings could only consist of a swarm of tiny satellites following Keplerian orbits around Saturn. At times, correspondence about his progress took a humorous turn, as when he wrote to his friend, the classical scholar Lewis Campbell (1830–1908), with a topical reference to the recently concluded eleven-month siege of Sebastopol by British, French and Turkish forces during the Crimean War:

> I have been battering away at Saturn, returning to the charge every now and then. I have effected several breaches in the solid ring, and now I am splash into the fluid one, amid a clash of symbols truly astounding. When I reappear it will be in the dusky ring, which is something like the state of the air supposing the siege of Sebastopol conducted from a forest of guns [160 km] one way, and [48,000 km] the other, and the shot never to stop, but go spinning away around in a circle, radius [270,000 km].[20]

Maxwell's own siege of Saturn's rings, using mathematics instead of cannonballs, ended with the publication, in 1859, of his famous essay 'On the Stability of the Motion of Saturn's Rings'.[21] But although there was never any doubt of its conclusions, direct confirmation came only in 1895, when the American astrophysicist James Keeler (1857–1900), then at the Allegheny Observatory in Pittsburgh, Pennsylvania, succeeded in observing the spectrum of the rings. By placing the slit of the spectroscope on different parts of the rings, he was able to show that the lines in the spectrum were tilted: velocities of particles making up the rings varied in the manner expected if they were moving in Keplerian orbits, with particles in the inner edge of the rings rotating faster than those at the outer edge, just as Maxwell had shown.

Kirkwood Gaps

But if the rings are indeed a swarm of satellites, what is the Cassini Division? Why should there exist a gap in this particular location? The American mathematical astronomer Daniel Kirkwood (1814–1895) of Indiana University discovered a clue in the late 1860s. Already, in 1866, he had shown that in the asteroid or minor planet belt between Mars and Jupiter gaps appear at distances where the orbital periods of asteroids are small integer fractions of Jupiter's orbital period, for example, 1/3, 2/5, 3/7, 1/2 and 3/5. Lying closer to the Sun and moving faster around it, an asteroid would pass Jupiter at the same point in its orbit time after time and be perturbed by the gravitational pull of the giant planet, rather in the manner of a child being pushed on a swing; the orbit and the swing both have a natural frequency, and – confining ourselves to the astronomical case – the planet produces a cumulative effect on the asteroid's motion, and pushes it out into a new, more stable orbit.

A year later, Kirkwood applied the same kind of reasoning to Saturn's ring. He found that the Cassini Division lay exactly where ring particles had periods of 1/2 with that of the innermost satellite then known, Mimas. This is an orbital resonance. The ring particles and Mimas exchange momentum and shift orbits (in the case of Mimas, the shift is very small) until the resonance no longer exists. The result is a gap in the rings.[22]

Other ring structures might be explained by means of resonances. Dawes, soon after the discovery of the C ring, had suspected a division between it and the inner edge of the B ring; it actually exists, and lies in the position of the 1:3 Mimas resonance. Kirkwood himself attempted to extend the resonance theory to Encke's Division, which he thought represented a sparse area in the rings rather than a true gap like Cassini's, while still later others attempted to extend it to other minor divisions seen from time to time.

Keeler's Magical Night

Though the fine structure of the rings appeared rather evanescently, depending on the instrument used and the quality of the 'seeing', that of at least one ring – the A ring – became rather clearer as a result of an extraordinary observation made on 7 January 1888, by Keeler, who at the time was one of the original members of the staff of the Lick Observatory on Mt Hamilton, California. That night was 'first light' for the 91-cm refractor, then the largest in the world, and Keeler would find the seeing superb, with crisply defined views even at 1,000×. Unfortunately, a week of frigid weather had frozen the immense iron dome's rotation mechanism, so objects could only be observed for periods of half an hour as they transited the dome's slit. As Saturn did so, Keeler recounted,

> he [that is, Saturn] presented probably the most glorious spectacle ever beheld. Not only was he shining with the brilliancy due to the great size of the objective [lens], but the minutest details of his surface were visible with wonderful distinctness. Most of these I had seen before with smaller instruments, but merely seeing an object when every nerve is strained, and even then with half a doubt as to its reality, is very different from seeing the same object glowing with abundance of light and visible at the first glance.[23]

Studying the image carefully, Keeler was able to make out a very narrow black line near the outer edge of Ring A. He estimated its position as 'a little less than one-fifth of the width of the ring from its outer edge', and described it as 'a mere spider's thread'. From this unprecedented view, Keeler was able to shed light on the somewhat-inconsistent and even contradictory descriptions of the structure of the outer ring given by other astronomers going back to

Encke – and indeed, what he describes agrees well with spacecraft views a century later:

> This line marked the beginning of a dark shade which extended inward, diminishing in intensity, near to the great black [Cassini] division. At its inner edge the ring was of nearly the same brightness as outside the fine division . . . It is easy to see how, with insufficient optical power, this system of shading could present the appearance of an indistinct line at about one third the width of the ring from its outer edge. The broad band alone would make it appear near the center of the ring, and the effect of the fine line, itself invisible, would be to displace the greatest apparent depth of shade in the direction of the outer edge.[24]

James E. Keeler's drawing of Saturn, with the 91-cm refractor of the Lick Observatory, showing it as it appeared on 7 January 1888, the first night of observations with the great telescope.

Cassini spacecraft image of the outer part of Ring A, showing the location of the Keeler Gap (not seen by Keeler but by *Voyager*) and the Encke Gap (not seen by Encke but by Keeler).

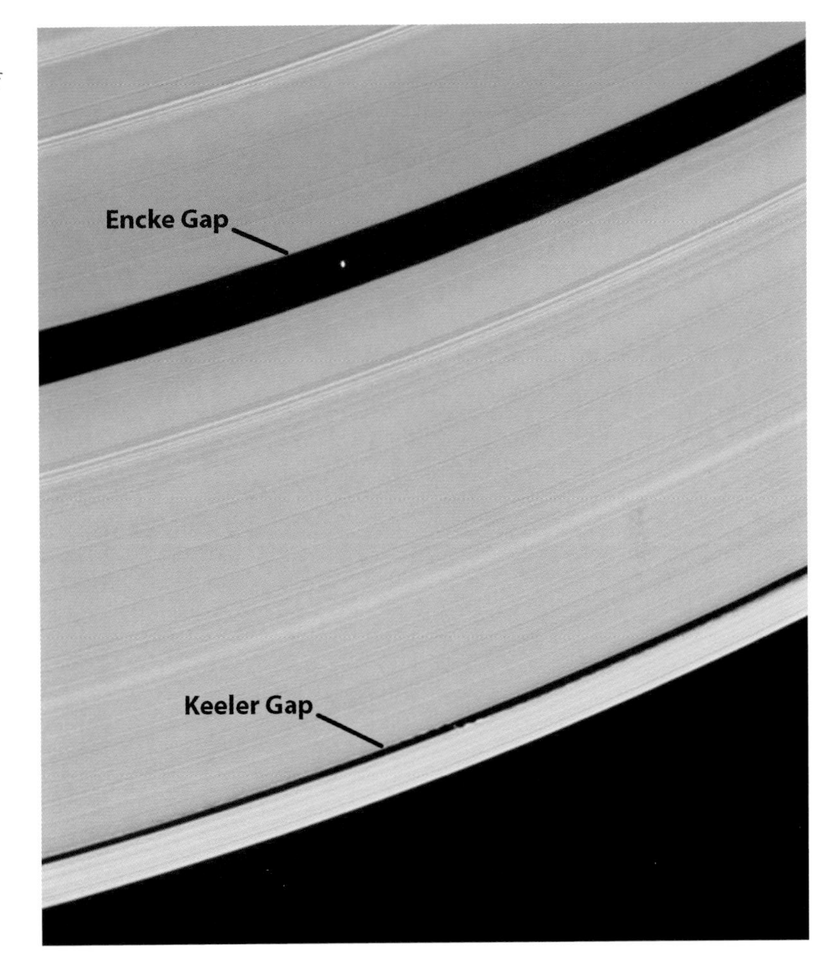

Indeed, Keeler's division, which has proved to be a true gap, 325 km wide, was finally imaged definitively by the *Pioneer 11* spacecraft in 1979, and then by the much more sophisticated *Voyager* I spacecraft the following year. NASA rather too hastily designated this the 'Encke Gap', the name officially adopted by the International Astronomical Union (IAU), even though it was discovered by Keeler, and Encke himself never saw it! The dusky band that Encke did see is often referred to as the 'Encke Minima' to distinguish it from the actual gap seen by Keeler. Meanwhile, the name 'Keeler Gap' was assigned by the IAU to another division, 35 km wide, and located a mere 270 km inside the outer edge of the A ring. It was first imaged by the *Voyager* spacecraft. In what follows, keep in mind that the names for these features are conventional, and have nothing to do with the actual history of their discoveries.

The Ball, Rings on Edge and More Moons

The rings are of almost whimsically infinitesimal mass compared to the planet they surround. According to the latest measurements, their mass is some 3/1,000,000,000ths that of the planet they encircle, so that were we to proportion our text to the relative masses of the rings and ball, the number of pages devoted to the rings would be three, but almost a billion would be devoted to the ball.

The rings will always, however, receive disproportionate attention relative to the other Saturnian phenomena, the ball itself and the moons. They distract the way the magician's handkerchief and magic wand does, riveting the sight to the exclusion of everything else. Glowing tributes to the rings' unique fascination are found in almost every book on astronomy, with the following, by William Frederick Denning (1848–1931), an amateur astronomer at Bristol, England, being particularly noble:

> The globe of Saturn is surrounded by a system of highly reflective rings, giving to the planet a character of form which finds no parallel among the other orbs of our system . . . Even old observers, who again and again return to the contemplation of this remarkable orb, confess they do so unwearyingly . . . The beautiful curving outline of the symmetrical image always retains its interest and is ever fresh.[25]

By comparison, the ball of the planet appears to be a rather muted version of Jupiter. The colours are butterscotch and brown, and there are belts and zones like those of Jupiter – and given names according to the same general scheme of nomenclature. Vivid colours are rare (though they do tend to come out in larger instruments), and there is certainly nothing like the Jovian Great Red Spot.

Saturn is a typical gas giant. Its rapid rotation, with a period of only 10 hours 33 minutes, gives it an oblate shape, that is, it is flattened at the poles and bulges at the equator. By convention, the diameters of the giant planets are measured from the tops of their cloud decks. Thus Saturn's equatorial diameter is 120,536 km, and through the poles it is 108,728 km. The difference, known as the oblateness, is, at 10 per cent, the greatest of any planet.

Nine Earths could fit across Saturn side by side. If Saturn were hollow, 764 Earths would be needed to fill it. However, owing to the lightness of the materials of which it is composed – as we now know, it is almost entirely hydrogen and helium – its density is less than that of water. It would, therefore, hypothetically float in a giant water-filled tub.

Again, because of the lightness of its being, despite its size, its mass is only 95 times that of the Earth; even so, this makes it a (distant) runner-up to Jupiter, with 318 Earth masses. Gravitationally, these two are the dominant bodies in the solar system, apart from the Sun. They contain over 90 per cent of the planetary mass, with the terrestrial planets, including the Earth, accounting for a paltry 4/10th of a per cent. (Mercury, Venus and Mars together make up not quite 1 Earth mass.) Important though the Earthlike planets seem to us, objectively speaking, the Earth and its brethren are little more than flotsam and jetsam.

What we see of Saturn, as with the other planets, is only the upper cloud deck, located in the troposphere (defined, as with the Earth, as the lowest level of the atmosphere, in which almost all the weather takes place). Above, the clouds are wispier, largely

The giant ringed world. Comparison of Saturn and Earth at the same scale. The Earth image was taken from *Apollo 17*.

transparent layers. Detailed knowledge of the atmosphere of Saturn, however, had to await the spacecraft era.

In the nineteenth century, it was widely believed, on the basis of their low densities, that the giant planets were hot and gaseous – worlds in arrested development, midway between a planet and a sun. An eloquent statement of this view was given by Richard A. Proctor in 1870. At least in the case of Jupiter, scene of picturesque and ever-changing clouds, he thought that such turbulent phenomena could hardly be powered by an energy source as feeble as the distant Sun:

> That enormous atmospheric envelope is loaded with vaporous masses by some influence exerted from beneath its level. Those disturbances which take place so rapidly and so frequently are the evidences of the action of forces enormously exceeding those which the Sun can by any possibility exert upon so distant a globe . . . We seem led to the conclusion that Jupiter is still a glowing mass, fluid probably throughout, still bubbling and seething with the intensity of the primeval fires, sending up continuously enormous masses of clouds, to be gathered into bands under the influence of the swift rotation of the giant planet.[26]

Saturn's most dramatic atmospheric upheavals are known as Great White Spots, or Great White Ovals. These periodic storms are unique to Saturn and become large enough and conspicuous enough to be seen even in small telescopes. The first record of a Great White Spot was made by Asaph Hall (1829–1907) at the U.S. Naval Observatory in Washington, DC, in December 1876. While studying the satellite Iapetus with the observatory's recently unveiled 66-cm refractor (then the largest in the world), Hall noticed a bright, well-defined spot in the equatorial zone of Saturn, and wrote at once to other astronomers to encourage them to add their observations. Unfortunately, he provided an incorrect ephemeris of when the spot would be visible. Instead of a period of 10 hours, 16 minutes and 0.4 seconds, adopted by William Herschel in 1794, Hall used a rotation period thirteen minutes longer – cited in almost all the textbooks of the day as supposedly that of Herschel though actually given by Laplace for the rotation period of the rings rather than the ball. Fortunately, the spot was so brilliantly white that it was easily recognized, even without an ephemeris, and proved to be long-lived by Saturnian standards; it was eventually seen by at least five other observers in America alone, and from nineteen observations of the spot across the central meridian of the planet over a four-week period, Hall derived a rotation period of 10 hours, 14 minutes and 23.8 seconds, with a probable error of only 2½ seconds.

Asaph Hall, who discovered the first Great White Spot of Saturn in December 1876.

Parenthetically, Hall's discovery that the often-repeated value of Saturn's rotation in the textbooks was erroneous made him generally sceptical of such information, and it was thus that he came to search for faint satellites of supposedly 'moonless' Mars the following year. He was successful, and in

August 1877 made the discovery for which he will always be remembered, that of Phobos and Deimos, the two dwarf satellites of Mars.

During the edgewise passages of the rings, the ball appears ringless for weeks in a small telescope, and for a few days even in the very largest instruments. Saturn then appears un-Saturnlike indeed, and reveals, without distraction, the kind of world it really is. It is a typical 'Jovian' planet – a smaller twin to Jupiter, though with the difference that in addition to atmospheric belts and zones, a thin and extremely dark line at the equator, the shadow of the rings on the ball, remains visible. (Always a reminder of the rings!) Even when nearly 'absent', the rings continue to take pride of place, for it is during the edgewise ring passages that some of the most penetrating insights into the nature of the rings have been revealed. In addition, the edgewise passages have often provided the best opportunity for new satellites to be recognized.

As mentioned earlier, the edgewise passages, in which the Sun and Earth cross the plane of the rings, occur at alternate intervals of 13 years 9 months and 15 years 9 months, during which the Earth passes either once or three times through the ring plane. When there are three ring-plane passages, the middle one occurs near Saturn's opposition and the other two near the quadratures (the points where the Saturn–Earth–Sun angle is 90°). Huygens had first worked out the circumstances of this during the triple-ring-passage year 1671–2.[27] When there is only one, the Earth and Saturn lie on opposite sides of the Sun at the time.

For reference, the following table lists some recent and future ring-plane passages:

Passages of the Earth and Sun through the ring-plane of Saturn

Year	Date	Object	Direction
1966	16 June	Sun	South to North
1966	28 October	Earth	North to South
1966	18 December	Earth	South to North
1980	3 March	Sun	South to North
1980	12 March	Earth	North to South
1980	23 July	Earth	South to North
1995	22 May	Earth	North to South
1995	10 August	Earth	South to North
1995	19 November	Sun	North to South
2009	11 August	Sun	South to North
2009	4 September	Earth	North to South
2025	23 March	Earth	South to North*
2025	6 May	Sun	North to South
2038	15 October	Earth	South to North
2039	1 April	Earth	North to South
2039	23 January	Sun	South to North
2039	1 July	Earth	North to South

*Since superior conjunction with the Sun occurs on 12 March 2025, this ring passage will be all but inaccessible to observation for Earth-based observers, since Saturn's elongation will be only 6° from the Sun.

During the edgewise passages, when Saturn appears 'reminiscent of a ball of yarn perforated by a knitting needle',the satellites appear like luminous beads upon a thread, and were even, somewhat naively, mistaken by some of the early observers for actual mountains standing in projection above the surface of the ring.[28] With the glare from the rings reduced, fainter satellites, out of reach at other times, may emerge, like stars in the sky during an air-raid blackout. Thus Cassini discovered Iapetus and Rhea when the rings were edgewise in 1671–2, and Dione and Tethys during the next edgewise presentation in 1684. Meanwhile, two new satellites came to light during the edgewise presentation of 1789.

An engraving showing William Herschel's 6.1-m (20-ft) focal length reflector, with a mirror of 47.5 cm (18.7 in.) in diameter, that William Herschel set up at Datchet, near Windsor. This was one of Herschel's most productive instruments, and used to discover two satellites of Saturn.

That year, William Herschel, with the great 6.1-m (20-ft) focal length reflector at his observatory at Slough, England, made a thorough examination of Saturn and its satellite and ring phenomena.[29] On 28 August, he entered the following observation in his logbook:

> Saturn with 5 stars in a line, very beautiful. The nearest of these five is probably a satellite, which has hitherto escaped

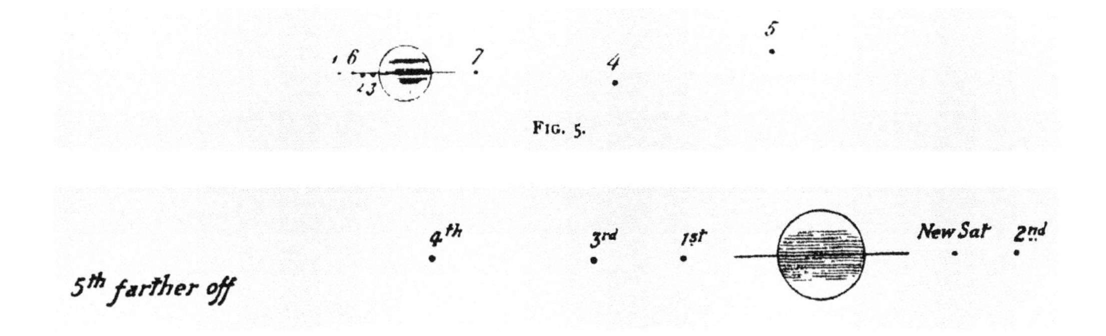

William Herschel's sketch of Saturn of 28 August 1789, using his great 12.2-m (40-ft) reflector to verify the sixth satellite (Enceladus), previously detected with the 6.1-m (20-ft) reflector.

Herschel's sketch of 20 October 1789, using the 6.1-m (20-ft) telescope, showing seven satellites lined up along the ring: 1 Tethys, 2 Dione, 3 Rhea, 4 Titan and 5 Iapetus, plus Herschel's two discoveries, 6 Enceladus and 7 Mimas.

> observation. It is less bright than the others. What makes me take it immediately for a satellite is its exactly ranging with the other four and with the ring.[30]

This satellite was that which his son Sir John Herschel later named Enceladus. William was tracking it on 8 September when he detected a still fainter one, which Sir John would name Mimas.

Yet another satellite – which would be the last Saturnian satellite discovered visually – came to light during the edgewise presentation of 1848. This is Hyperion, which was first recognized by the Director of the Harvard College Observatory, William Cranch Bond (1789–1859), with the 38-cm Merz and Mahler refractor, and then independently, just a few nights later, by William Lassell with the 61-cm equatorial at Starfield (near Liverpool). Since it orbits between Titan and Iapetus, it is often far from Saturn, so despite its faintness (fourteenth magnitude), much of the time it is well clear of the glare of Saturn. Nevertheless, few amateurs have ever seen it, mainly because they have not taken the trouble of working out where to look.

By using the satellites as they passed in front of or behind the rings as micrometers during the edgewise presentation in 1789, William Herschel attempted to estimate the rings' thickness. He put this at no greater than 320–480 km. This, however, was far too large, and successive observers would pare down this value until

– with the definitive results from spacecraft – we can now say with certainty that the rings' average thickness is no more than 10 m, although, in parts, they may reach several kilometres owing to the presence of vertical relief produced by bending waves. Were Saturn reduced to the scale of a basketball, the rings would have roughly 1/250th the thickness of a human hair.

Curiously, the visibility of the rings on opposite sides of the globe is often unequal. When attempting to determine the visibility of the rings under these conditions, it is important to exercise caution since – as the American astronomer Edward Emerson Barnard (1857–1923) discovered on observing Earth's passage through the rings in October 1891 with the 91-cm refractor of Lick Observatory – what might seem to be occasional glimpses of the rings may be nothing more than after-images of the dark shadow of the rings on the globe.

During periods when the Sun and Earth are on opposite sides of the ring plane, the dark face of the rings is in view, and any light has obviously either been reflected from the globe or has been filtered through the rings. Faintly luminous patches are then visible along the rings, known among nineteenth-century astronomers as 'knots' or 'condensations'. Herschel, of course, had noticed them in 1789, and scrutinized them closely. Though most of them were found to be satellites superimposed on the thread of the ring, he was left with a residual set of almost fifty observations, chiefly recorded in October–December 1789, which did not fit the position of any known satellite. The knots and condensations were recorded by all the leading Saturn observers of the nineteenth century, and were long regarded as mysterious, but Barnard, in a crucial series of observations with the 1.02-m refractor of Yerkes Observatory, during Earth's passage through the ring plane in January 1908, showed that they lined up precisely with the positions of the C ring and the Cassini Division.[31] From this, Barnard concluded that neither the C ring nor the Cassini Division is entirely devoid of particles, which

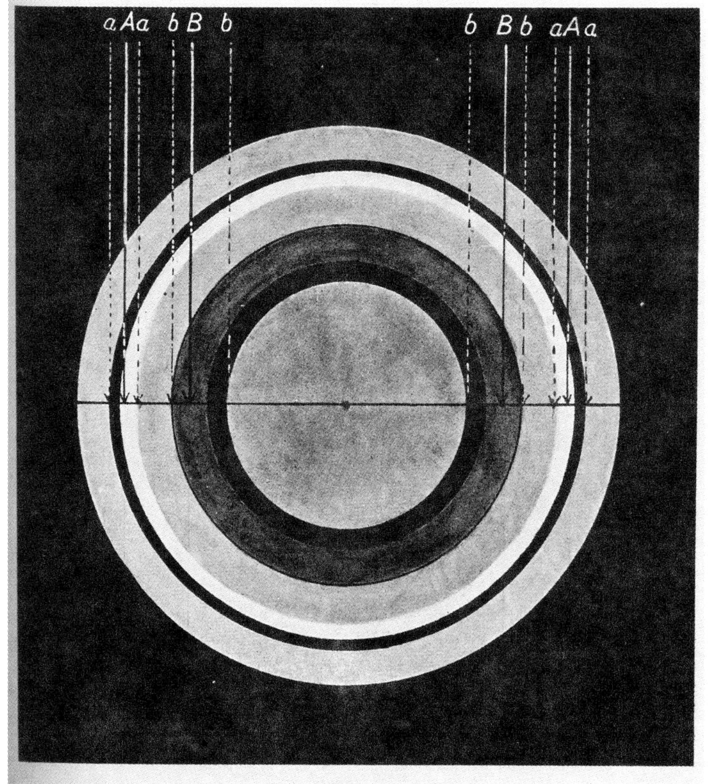

Above: Saturn by E. E. Barnard with the 1.02-m refractor of Yerkes Observatory, 25 November 1907. *Below*: the lines A, B, B, A show where the centres of the condensations fall, the dotted lines a, a, b, b, b, b, a, a show their limits, and relate them to structures in the rings.

we now know from spacecraft observations to be quite correct. The 'condensations' are due to the fact that, when we look at the dark side of the rings, although sunlight is almost completely back-scattered by the A and B rings and virtually none of it gets through, the sparser material of the C ring and the Cassini Division allows some forward-scattering, making these areas faintly luminous. This, by the way, corresponds to the views we have now got used to in looking back at Saturn from a fly-by or orbiting spacecraft.

A Bevy of Moons

Saturn has a retinue of moons second only to Jupiter. Huygens's large moon Titan circles Saturn once every sixteen days at a distance of 1,221,800 km from the planet's centre. With a diameter of 5,149 km, it is larger than the planet Mercury, and ranks second among all the moons in the solar system behind only Jupiter's Ganymede (5,162 km). It is also the only one with a substantial atmosphere, discovered by Gerard Peter Kuiper (1906–1973) with a spectrograph attached to the 2.1-m reflector of the McDonald Observatory, in Texas, from the spectral signature of methane.

Among Cassini's discoveries, Iapetus, 1,471 km in diameter, orbits at a distance of 3,560,900 km from Saturn, and has a period of 79 days. It is remarkable for being, as Cassini himself noted, nearly two magnitudes brighter when west of Saturn than when east, which implies that one hemisphere is covered with more reflective material than the other. Cassini's other satellites, Tethys, Dione and Rhea, range in distance between 294,700 and 527,100 km, with Rhea being the largest with a diameter of 1,529 km; the others are just over 1,000 km.

Inside Tethys, we have Enceladus, with a diameter of 499 km, and travelling in an orbit 238,100 km from the centre of Saturn, with a period of 1.37 days; and then Mimas, 397 km in diameter,

Cassini image showing Saturn's rings by forward scattered light. The Sun is behind Saturn, and the sunlight is filtering through the relatively sparser Ring C, outer Cassini Division, the A ring, and the F ring.

and moving in an orbit 185,600 km from the centre of Saturn, with a period of 0.942 days. Even in large telescopes, these two appear much as they did when William Herschel discovered them, but as we now know from spacecraft observations, they are worlds in their own right, with Enceladus, in particular, having emerged as perhaps the most promising place beyond Earth to search for life (see Chapter Six). Hyperion is an irregularly shaped body, and as it travels round Saturn, it tumbles wildly; its rotation is a classic example of chaotic motion, changing without any predictable pattern.

Well outside all the rest lies Phoebe, the first satellite in the solar system to be discovered photographically, by Harvard astronomer William H. Pickering (1858–1938) in 1898. It is located at a distance of 12,944,300 km from Saturn, with a period of revolution of 550

Hubble Space Telescope image of the ball of Saturn, displaying its multicoloured bands during the April 1996 edgewise apparition. The shadow on the ball is cast by Titan, visible at upper left, while in a bunch just to the right of the ring are Mimas, Tethys, Janus and Enceladus.

Titan and Saturn photographed by *Cassini*, 6 May 2012.

days, and travels through its orbit in the opposite sense – or retrograde – to the other satellites, with the plane of its orbit sharply tilted to that of Saturn by 150°. It is not quite but fairly round in shape, with a diameter of about 214 km, and has left in its wake a faint outer ring, known as the Phoebe ring, discovered by the Spitzer Infrared Space Telescope in 2009, and believed to consist of dust particles ejected from Phoebe's surface by meteorite impacts.

Pickering, in 1904, reported the discovery of yet another faint moon, which he called Themis. The images which led to this identification were, however, presumably those of faint field stars or possibly an asteroid near its stationary point. In any case, Themis does not exist.

No more satellites were reported until 1966, when the edgewise ring presentation provided excellent conditions for searchers. The French astronomer Audouin Dollfus (1924–2010), suspecting there might be further inner satellites by the failure of the divisions in the rings to line up exactly with known resonance positions, used the 61-cm refractor at the Pic du Midi Observatory, equipped with a special filter to cut down the planet's glare, to record the image of an inner tenth satellite, which he called Janus. Later, a reanalysis of these observations by Stephen Larson and John W. Fountain of the University of Arizona found that they better fitted two (or more) satellites rather than just one, and spacecraft showed this to be the case: there are two co-orbital satellites, which have received the names Janus and Epimetheus. Janus measures 220 by 160 km, with Epimetheus slightly smaller at 140 by 100 km. They both lie at a distance of about 151,400 km from Saturn, and have nearly identical periods of revolution, about 16 hours 40 minutes. The distance between them is actually less than the sum of their diameters, but because their periods differ by about half a minute, they encounter one another every four years. When they do, they swap orbits, with the one that had been slightly closer to Saturn

Epimetheus and Janus, imaged from *Cassini* on 20 March 2006, two months after they swapped orbits. Though they appear next to each other, they were actually some 40,000 km apart: the spacecraft being 452,000 km from Epimetheus and 492,000 km from Janus at the time.

moving farther out and vice versa. They are the outermost of five small satellites that seem to have formed in close association with the rings, intimately involved with their structure and known as 'ringmoons'. Needless to say, none of these is remotely within the reach of amateur observers.

On the eve of the spacecraft era, in 1979–80, three more small moons were found by Earth-based observers at Lagrangian points – gravitationally stable locations occupying equilateral points about Saturn relative to other moons: Helene shares an orbit with Dione, and Telesto and Calypso with Tethys. These were the last Saturnian moons to be discovered from Earth. The rest are only the size of asteroids, which is presumably what they are, and were revealed by spacecraft.

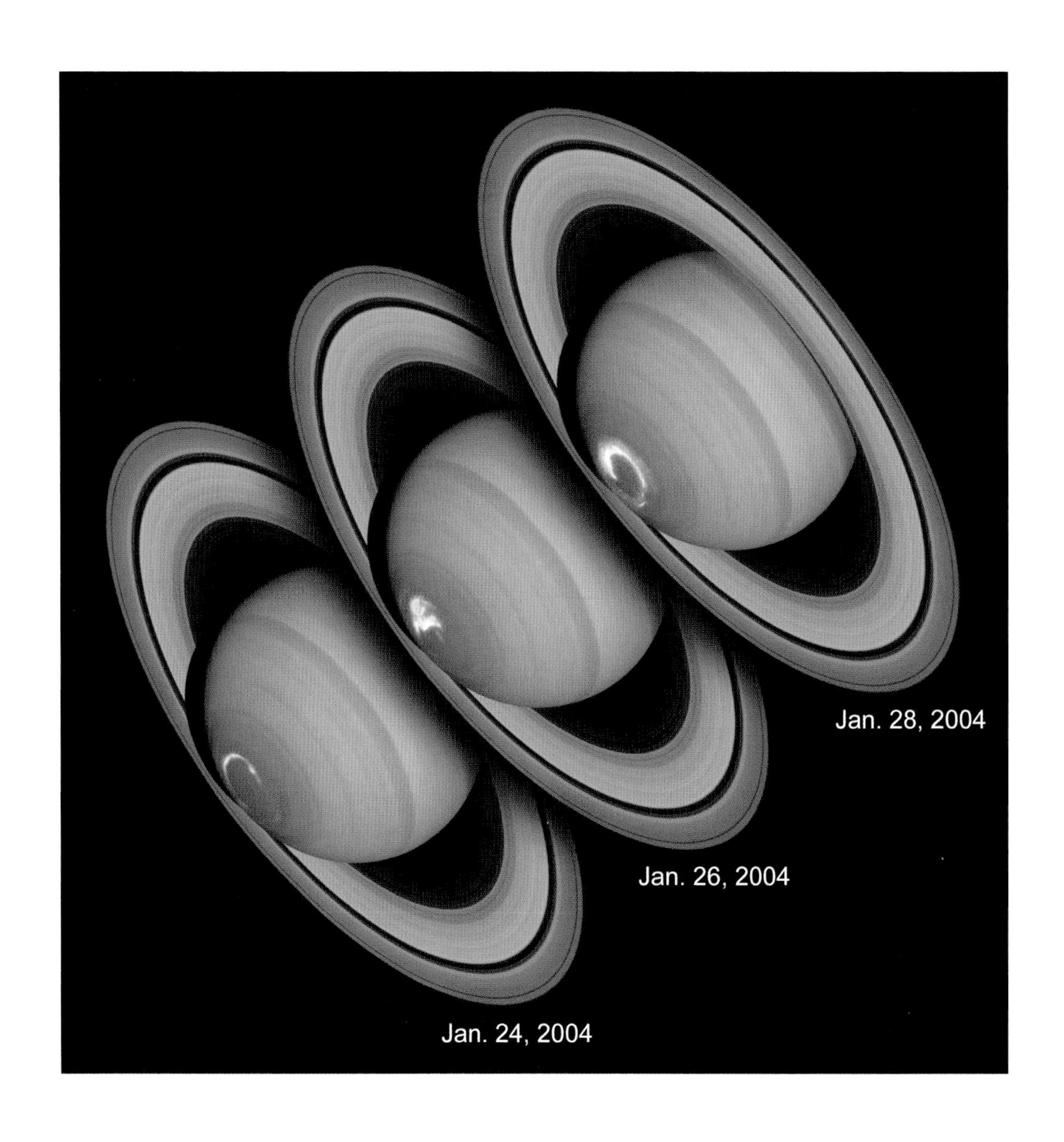
Jan. 28, 2004
Jan. 26, 2004
Jan. 24, 2004

THREE

Saturn in Depth

Nineteenth-century astronomers believed that the giant planets, including Saturn, were hot and gaseous – worlds that fell just short of being full-fledged suns. They also guessed that the disturbances observed in the clouds, especially on Jupiter, occurred so rapidly and with such frequency that they could only be explained by assuming that these planets generated some of their own heat. As Proctor said, 'We seem led to the conclusion that Jupiter is still a glowing mass, fluid probably throughout, still bubbling and seething with the intensity of the primeval fires.'[1]

This was an inference from what was observable in the troposphere of Jupiter, which (as on Earth) is the layer in which most of the 'weather' occurs. On Saturn, this region – where the pressure is about 1 bar, or about the same as the surface pressure on the Earth – is where the visible cloud phenomena, which were studied by classical observers, occur.

A montage in which false-colour images of ultraviolet auroral activity taken with the Imaging Spectrograph of the Hubble Space Telescope on 24, 26 and 28 January 2004 have been superimposed on a visible-light image taken with the Advanced Camera for Surveys on 22 March 2004.

At higher altitudes, where the pressure is less than about 100 mbar, the atmosphere is thin and stably stratified, and forms the Saturnian stratosphere. This region is dominated by photochemical processes and bombardment by energetic particles, especially in the auroral zones, to produce a haze of stratospheric aerosols (for example, linear chains (polyacetylenes) or polycyclic aromatic hydrocarbons produced from methane).

As on Earth, auroras are produced by Saturn's magnetic field, which is thought to be generated by a zone of electrically conducting fluid deep inside the interior (probably in the metallic hydrogen layer; see below). The magnetic field creates a zone, called the magnetosphere, into which magnetic fields from the Sun are unable to penetrate, and as in the Van Allen Radiation Belts of Earth, Saturn traps charged particles (ions) within this zone. Charged particles channelled along the magnetic field lines into the upper atmosphere produce Saturnian auroras, which have been observed in a narrow latitudinal band between 78° and 81.5° in both hemispheres.

Though together hydrogen and helium, the most common elements in the universe, make up most of the bulk of the giant planets – well over 99 per cent – other gases, such as methane, ammonia and ammonium hydrosulphide, are present, and account for the colourful cloud features in their tropospheres. The levels at which clouds of different composition condense into droplets or solid flakes are determined by the temperature and pressure: at a pressure of 1 bar, methane condenses at about 75 kelvin (–198°C), ammonia at 150 kelvin (–123°C), ammonium sulphide at 200 kelvin (–73°C) and water at 273 kelvin (0°C). This explains why characteristic cloud layers form at successive depths. Methane condenses only at the very cold temperatures found in Saturn's stratosphere, while the tropospheric cloud layers are as follows:

1. an upper cloud deck of white ammonia clouds;
2. a clear zone;
3. a layer of brownish ammonium sulphide clouds;
4. a layer of water-ice clouds; and
5. a solution of ammonia and water.

Similar cloud layers are found on Jupiter, though given Jupiter's greater gravity and higher helium/hydrogen ratio, the clouds are compressed within a much narrower vertical range than on Saturn.

Nomenclature of the belts and zones of Saturn.

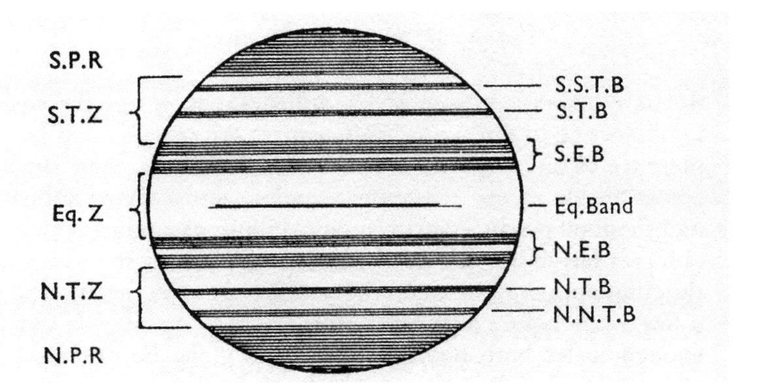

On Jupiter, ammonia clouds, some 10 km thick, lie some 30 km above a layer of ammonium-sulphide clouds, and 50 km above a layer of water-ice clouds. At Saturn, the ammonia clouds occur at a depth of some 100 km, the ammonium hydrosulphide clouds at 200 km and the water-ice clouds at 275 km. By contrast, the two outer gas giants, Uranus and Neptune, are far enough from the Sun for methane to freeze out, so methane clouds, rather than ammonia clouds, form their upper cloud decks – which explains why these remote planets appear bluish-green rather than whitish-yellow in the telescope.

Saturn's clouds show the same pattern of global circulation as Jupiter's, being drawn by the rapid rotation into a series of darker and lighter bands which are conventionally known as belts and zones, respectively. Saturn's features appear less distinct, however, mainly because of the greater vertical depth of Saturn's atmosphere compared to Jupiter's: there is simply more gas present above the clouds, and a thicker layer of aerosol haze to gaze through. The belts and zones are named by analogy to those on Jupiter: thus there is a bright Equatorial Zone, bordered on each side by dark belts; the North Equatorial Belt and the South Equatorial Belt, and so on, as shown in the diagram.

Since we are creatures that live on the surface of our world, and breathe the air around us, the troposphere is our element, and naturally we are apt to treat the cloud decks of these planets as if they contain most of what is of interest. In terms of visible features, that is certainly the case – the troposphere is where the 'weather' is. However, it is important to emphasize that these gaseous planets are very unlike the Earth; they have no solid surface on which it would be possible to stand, and the visible atmosphere makes up, as with the Earth itself, only a fraction of the entire mass. Most of the bulk of these planets lies in the deep interior, which we cannot see directly.

The table below shows the measured proportion of gases in the atmosphere of Saturn compared to those of the other giant planets and the Sun. Note that in each case hydrogen and helium overwhelmingly dominate, making up some 98 or 99 per cent of the atmosphere of each planet.

Gases detected in the atmospheres of the giant planets and the Sun, showing the measured proportion (%) made up by each

	Sun	Jupiter	Saturn	Uranus	Neptune
Hydrogen (H_2)	84	86.4	97	83	79
Helium (He)	16	13.6	3	15	18
Water (H_2O)	0.15	(0.1)	–	–	–
Methane (CH_3)	0.07	0.21	0.2	2	3
Ammonia (NH_3)	0.02	0.07	0.03	–	–
Hydrogen sulphide (H_2S)	0.003	0.008	–	–	–

Note: All figures are uncertain in the least significant figure; dashes indicate no data.

Saturn: Tropospheric Wind Patterns

The belts and zones on Saturn are related, as on Jupiter, to a series of zonal jet currents, which began to be tracked by visual observers over a century ago and have now been definitively studied with the Hubble

Space Telescope and spacecraft. It appears that in general features have been remarkably stable for at least a century. On Saturn, the zonal jets are quite symmetrical about the equator, and show a predominantly eastward flow; also, the mid-latitude eastward jets on Saturn are faster than those on Jupiter. The most extreme case is the powerful equatorial jet, where the wind speeds are 1,800 km/h – two-thirds of the speed of sound at that level of Saturn's atmosphere. Observations with instruments able to penetrate the haze and look into the troposphere have revealed that the width of the belts and zones varies with altitude: the darker belts are narrower and lie in the positions of the rapid jets; the brighter zones are wider and coincide with slower jets and may even be stationary relative to the general rotation of the planet. At the boundaries between jets, breaking waves occur, and wind shear gives rise to swirling vortices which may collide, merge and disappear. Most of these last only a matter of weeks, but some are longer-lived. For example, a one-third-sized version of Jupiter's Great Red Spot was recorded in Saturn's high southern latitudes (55°S) by *Voyager* 1 in 1980. It continued to be tracked during the *Voyager* 2 encounter until September 1981.

The *Voyagers* also discovered a bizarre hexagonal pattern around Saturn's north polar vortex, of which each side of the hexagon measures 13,800 km; the period of rotation of the vortex is 10 hours 39 minutes and 24 seconds. The hexagonal structure is thought to reflect a standing wave pattern in the atmosphere. (Similar shapes have been produced in laboratory experiments through the differential rotation of fluids.) Though there is also a south polar vortex, it does not give rise to hexagonal structure.

Colours

In general, the colours on Saturn are muted butterscotch and brown. However, there are subtle seasonal changes. Observers of the British Astronomical Association (BAA) Saturn Section, for instance, noted that from 1947 to 1950, a period which saw late summer in Saturn's southern hemisphere and late winter in the northern, the southern hemisphere appeared 'warmer' and the northern 'cooler' in colour (brownish-tinged versus bluish-tinged).

Observing visually with the 1.06-m Cassegrain at Pic du Midi and the 1.02-m refractor at Yerkes Observatory, Wisconsin, in 1992–3, the author found the entire southern hemisphere above the closing ring appeared so intensely greenish-blue as to be described as turquoise.[2] However, within two years of the passage of the Sun north to south through the ring plane in 1995, the southern hemisphere appeared noticeably browner, and it was the northern hemisphere's turn to become greenish-blue.

It appears that the seasonal colour changes are owing to a combination of the lower solar ultraviolet flux in autumn and winter, when there are fewer hours of daylight, and the direct effect of blocking by the shadow of the rings. Under these conditions, the usual aerosols do not form, and the upper atmosphere is mostly clear, allowing Rayleigh scattering of sunlight (scattering of short wavelengths) like that which makes the cloudless sky on Earth blue. In addition, methane in the atmosphere is effective in absorbing red rays from the solar spectrum. Since on its distant, colder cousins, Uranus and Neptune, the ammonia and water clouds form much deeper than at Jupiter and Saturn, their upper levels show the same blue colour of a cloudless sky. In Saturn's blue hemisphere we see then, in the words of *Cassini* Imaging Team leader Carolyn Porco, 'a slice of Neptune's atmosphere spliced onto Saturn'.[3]

The shadow of the rings on the globe has a marked effect on the planet's weather. Not only is the temperature in the rings' shadow reduced, but the rings' shadow affects the ionized part of Saturn's upper atmosphere, known as the ionosphere and located at a distance of 300 to 5,000 km above the level of the visible cloud deck. By cutting down the amount of ultraviolet radiation coming from the Sun, fewer particles become ionized (acquire charge by losing

The hexagonal pattern round the north polar vortex, of which each side of the hexagon measures 13,800 km. There is no similar feature round the south pole.

Images used to create this *Cassini* image were taken on 29 July 2013, and show the complex ring shadows projecting against the ball of the planet. Note the bluish colouring below the ring shadows, which indicates that the southern hemisphere is entering winter.

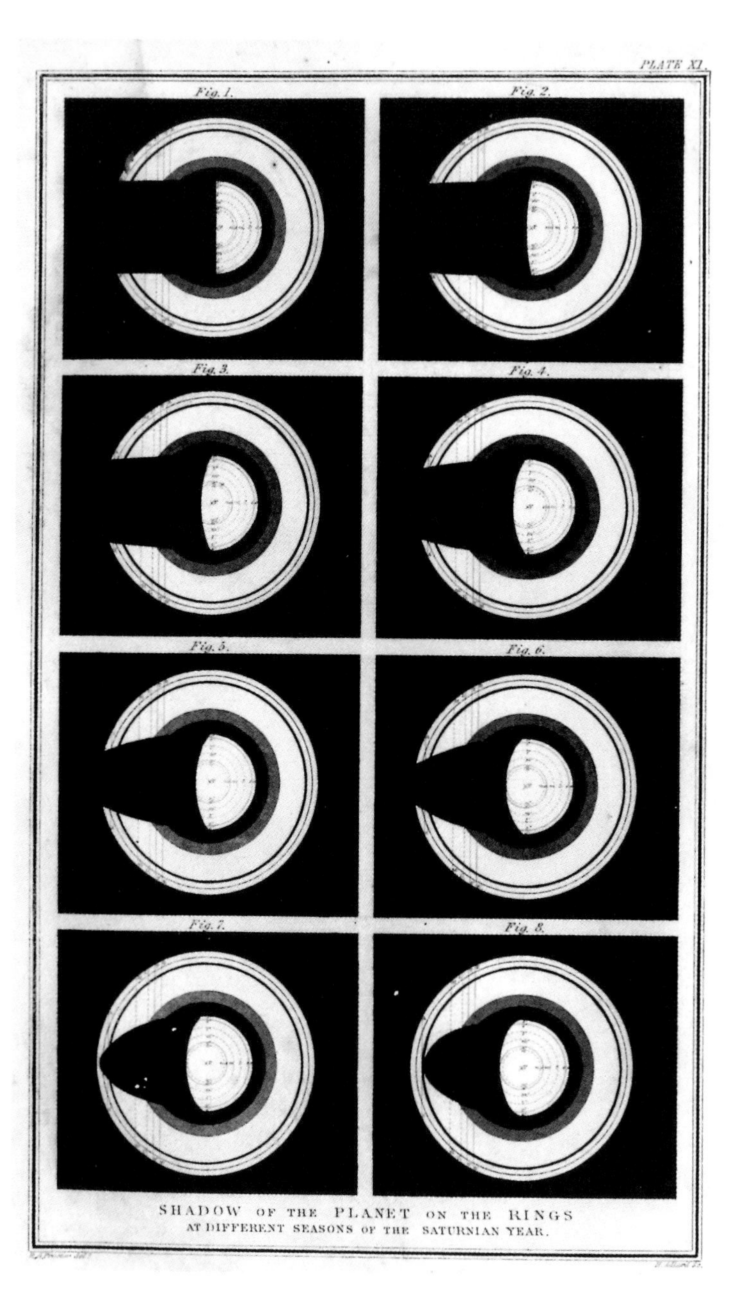

Aspects of the shadow of the globe on the rings, at different Saturnian seasons. Upper left is the situation at equinox, the lower right at solstice.

Shadow of rings on the globe, at different seasons; upper left, equinox; lower right, solstice.

electrons). During Saturn's equinoxes, when the Sun passes through the ring plane and stands directly over the equator, the Sun is effectively 'turned off' over the rings. For about four days, night falls over the entire ring system, and the temperature of the rings drops precipitately – according to measurements by the *Cassini* spacecraft during the ring-plane passage on 11 August 2009, the temperature fell to 43 kelvin (–230°C).

The Great White Spots: Saturn's Signature Atmospheric Phenomenon

The most dramatic atmospheric upheavals are known as Great White Spots, or Great White Ovals. After the first Great White Spot (GWS) discovered by Asaph Hall in 1876, the next was not observed until 1903, and it was not in the equatorial zone but in the north temperate zone (at latitude 36°N). It was first noted in mid-June by Barnard with the 1.02-m refractor at Yerkes Observatory. He was quite surprised by it, as he had never previously succeeded in seeing anything of the sort.[4] It was independently discovered by the well-known British amateur William Frederick Denning with a 25-cm reflector at Bristol, and by the Spanish astronomer José Comas y Solá (1868–1937) with a 15-cm refractor at Barcelona. It remained visible for several months, and was associated with an outbreak of various small white and dark spots whose rotation periods were all about 10 hours 38 minutes.

In August 1933, another GWS appeared in the Equatorial Zone. The discoverer was W. T. ('Will') Hay (1888–1949), a British stage and screen comedian best known for his comic schoolmaster sketch (and listed as the third-highest-grossing star at the British box office in 1938, behind George Formby and Gracie Fields). A keen amateur astronomer, Hay noted the spot on 3 August with a 15-cm refractor at his private observatory in Norbury, a suburb of London. 'It was almost by accident that I happened to look at Saturn the other night,' he remarked at the time.

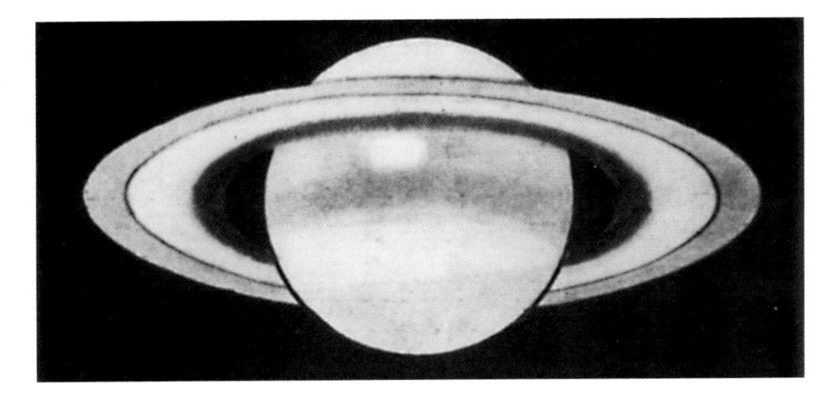

W. T. Hay's discovery drawing of the equatorial white spot, 3 August 1933.

> Like other astronomers, I often take a look at Saturn because, with its rings, it is the most beautiful of all the planets. When I saw the spot I admit I felt excited. I telephoned Dr W. H. Steavenson [of the British Astronomical Association] at once – just to make sure I was not 'seeing things'. If I had not seen it the spot would have been observed sooner or later.[5]

So it happened. The GWS was independently discovered two nights later by an astronomer at the U.S. Naval Observatory in Washington, the first of several independent discoverers. The rotation period was 10 hours, 14 minutes and 24.2 seconds, almost identical to that of Hall's spot, and with a second spot, smaller but more sharply defined, trailing some 75° behind in longitude, it underwent the same kind of explosive growth and expansion that the GWS of 1876 had done.

The next GWS, located in high latitudes (60°N), was noted on 31 March 1960 by the South African amateur astronomer J. H. Botham with a 15-cm refractor, and independently almost a month later by the French astronomer Audouin Dollfus with the 61-cm refractor at the Pic du Midi Observatory. The rotation period, as published by Dollfus, was about 10 hours 40½ minutes.

The fact that a GWS had appeared once every thirty years was noteworthy, and in 1989 led the Spanish planetary astronomer Agustín Sánchez-Lavega to suggest, on purely empirical grounds, that a new one would soon break out. This proved to be the case. On 24 September 1990, amateur astronomer Stuart Wilber, using a 25-cm reflector at Las Cruces, New Mexico, detected a small bright spot in the Equatorial Zone, which within the next few days was viewed by the author in only a 60-mm refractor; over the following week, the spot proceeded to lengthen into an oval cloud some 15,000 km long. By late October, whitish clouds encircled the whole globe, separated by a dark band along the border of where they were being sheared apart by faster wind currents just north and south of the equator; also beautiful festoons appeared along the northern edge. This was the view captured by the Hubble Space Telescope in early November. (Unfortunately, this was during its spherically aberrated state, before corrective optics were installed.) Sporadic outbreaks of whitish spots in the Equatorial Zone continued for the next six years, with an especially noteworthy occurrence in mid-July 1994, beginning with a small whitish spot preceded by a dark column. Over the next several weeks, it developed into a prominent oval feature easily visible in amateur telescopes.

The appearance of the 1990 GWS reinforced the view that this was a seasonal phenomenon, concentrated in the summer of Saturn's northern hemisphere. In that case, the next – and sixth – outbreak ought not to occur until sometime about 2020.

The Great White Storm of 2010

In fact – and luckily for us – the next GWS occurred early, with the *Cassini* spacecraft enjoying a ringside seat in orbit around Saturn. Beginning as a giant thunderstorm about one hundred times more intense than a typical thunderstorm on Earth, with intense lightning and cloud disturbances, it gave rise to a spectacular expansion of dense cumulus clouds of ammonia crystals that quickly encircled the planet and remained visible for six months.

The 2010 storm began with a white spot at latitude 37°.7 N that was first noticed by vigilant amateur astronomers Sadegh Gomizadegh in Iran and Teruki Kumamoari in Japan on charge coupled device (CCD) images taken on 8 and 9 December. A reanalysis of earlier images found that a small white spot had already appeared as early as 5 December. On that date, *Cassini*'s Radio Plasma and Wave Science experiment captured powerful radio emissions from intense lightning activity here, while the *Cassini* Imaging Science Subsystem recorded a 1,000-km-wide white spot in the same position. Though *Cassini* was unable to obtain another image until 22 December, amateurs kept it under intense surveillance, as the head of the disturbance proceeded to expand eastwards and develop a tail. By 10 December, it had grown to a width of 8,000 km, and was easily visible in small telescopes as a brilliant, whitish patch. Its behaviour followed that of previous storms. Within 55 days of its initial appearance, the head, blown westwards by winds of 30 ms, had caught up with the end of the tail, encircling the entire planet in a beautiful filigree of dense cumulus ammonia-ice clouds. These clouds towered high into the usual haze layer that overlies the planet's upper cloud deck, powered by hot moist buoyant gas rising, as if propelled by a coiled spring, from the deeper atmosphere.

The 2010 storm occurred in northern mid-latitudes, like those of 1903 and 1960. Indeed, it now appears that the storms occur alternately between mid-latitudes and the equator, so that the period between successive mid-latitude or equatorial outbursts is approximately sixty years. On the basis of numerical modelling, the California Institute of Technology's (Caltech) Andrew Ingersoll and graduate student Cheng Li have explained this quasi-periodic occurrence by the action of a water-loading mechanism: moist convection is suppressed for decades, owing to the relatively large molecular weight of water in Saturn's predominantly hydrogen–helium atmosphere.[6] When the water rains out, the upper

The Great Northern Storm. The tail of white ammonia cumulus clouds is wrapping itself round the planet, in this image taken 25 February 2011, or approximately twelve weeks after the storm began.

atmosphere becomes lighter, and remains so until radiative cooling (in the troposphere above the cloud base) overrides the suppressed convection; but because the upper atmosphere is so cold and massive, the radiative cooling occurs very slowly, and takes twenty to thirty years to trigger another storm. The precipitate onset occurs as buoyant, warm, moist air rises rapidly from below, with the dramatic results just described – the onset of a thunderstorm on a truly gigantic scale, as large as the entire Earth.

According to Ingersoll and Li's calculations, it appears that the amount of water vapour in the atmosphere must reach a critical level for the planet-encircling storms to develop. Saturn has water vapour above the critical level – the mixing level of water vapour to hydrogen and helium is 1.1 per cent; Jupiter is relatively water-depleted compared to Saturn, with a mixing level half as great, which explains, presumably, why it fails to develop planet-encircling storms.

Inside Story

The nineteenth-century idea of the planets being hot, fluid masses within was a lucky guess, and purely impressionistic. It was only in the early 1920s that the first rigorous mathematical models of the planets' interiors began to be developed, by the Cambridge mathematician and pioneering geophysicist Harold Jeffreys (1891–1989).

Jeffreys started with the best available information at the time about the size, shape, mass and rate of rotation of the planets, and based his models on varying assumptions about their assumed composition in an effort to get a good match. This is still what is done today, though, of course, the fundamental information has now been provided by spacecraft instead of ground-based observations. In addition, we now have much more detailed knowledge

A series of images showing the development of the Great Northern Storm.

of the compositions of the giant planets and of the properties of bulk matter – such as the equation of state of dense hydrogen–helium mixtures in the case of Jupiter and Saturn. Present models are accordingly much better constrained than in Jeffreys's day, but at least he was on the right track about some of the main features.[7]

According to his analysis, the giant planets were not half-fledged suns as had been supposed. Instead they had cold atmospheres and solid interiors, and were composed of materials of low density compared to terrestrial rocks. In January 1926, Jeffreys outlined his theory of the 'Physical Condition of the Four Outer Planets' at a meeting of the BAA, which included many of the leading amateur observers of the planets. He admitted that he felt 'rather like Daniel intruding into a den of lions', since he personally had never seen any of the planets through a telescope of more than 5 cm aperture.[8] A number of objections were raised; one celebrated observer insisted that on the basis of his own observations, the great atmospheric disturbances observed on Jupiter could only be explained by assuming an analogy between that planet and the Sun, while another offered a hopeful compromise – perhaps the interior of Jupiter was hot and molten though the surface was very cold, so that great volcanic eruptions might give rise to atmospheric disturbances (such as Jupiter's Great Red Spot) and even the occasional ejection of a short-period comet.

In the end, Jeffreys prevailed. His ideas about the interiors of the giant planets proved to be basically sound, and continued to influence later, more refined, models of the interiors of Saturn and the other giant planets, which have now relied heavily on spacecraft data. Indeed, by the time the *Cassini* spacecraft went into orbit round Saturn, in July 2004, the models had become quite detailed, rigorously constrained and consistent.

It was assumed that below the troposphere and the visible clouds, the structure became rather blandly uniform and homogeneous, as determined by the properties of hydrogen and helium at increasing depths and pressures. At a depth of about 1,000 km, hydrogen and helium become liquid. Accordingly, at this depth, a liquid hydrogen layer would form and sit on a layer of heavier liquid helium. Under these conditions, the molecules of hydrogen still hold together; electrons are not able to move about freely, and so hydrogen behaves as an insulator – that is, as a poor conductor of electricity. At still greater depths and pressures, the molecules of hydrogen break apart to form individual atoms with the electrons now freely flowing between them. The hydrogen has now become a conductor of electricity, which is the definition of a metal. The transition from fluid hydrogen to metallic hydrogen – realized on Earth only under extreme conditions created using lasers in the laboratory[9] – occurs at some 30,000 km below Saturn's cloud tops, where the pressures reach some 3 million bars. At still greater depths, near the very centre – where the temperature is 11,700°C, or equal to that of the photosphere of the Sun, and the pressures reach 50 million bars – a massive core of nickel–iron and rock might reside. Jupiter was supposed to be similarly constructed, though, of course, with even greater temperatures and pressures (up to 100 million bars at the centre).

The expectation for both planets was that below the upper 'weather' layer, everything would be fairly uniform, producing a fairly spherical and symmetric 'gravity field'. However, a more complicated picture has begun to emerge with the arrival of the *Juno* spacecraft in orbit around Jupiter in July 2016, with implications for Saturn as well.[10] Instead of being uniform and homogeneous within, Jupiter, at least, appears to be rather poorly mixed. Gravity mapping by *Juno*'s Gravity Science instrument shows that the gravity field is shaped differently in the northern hemisphere than in the southern hemisphere, and corresponds to the difference between the zones and belts in the two hemispheres. This asymmetry may mean that the thick lower layers contain deep atmospheric flows transporting mass at different rates in the north than in the south. It also appears that the belts and zones of Jupiter, which are produced by alternating atmospheric jet streams, extend far deeper than had been believed – to some 3,000 km below the cloud tops. This was unexpected, since it was thought that distinctly rotating sections of the atmosphere at such depths would dissipate, mix or drag each other to uniform speeds. Still greater depths appear to be largely unaffected by the alternating jet streams, and the bulk of the planet – 99 per cent by mass – rotates as one solid body from pole to pole. Thus the alternating movements of the upper belts and zones decouple at about the 3,000-km level.[11] Because Saturn's wind patterns are different from those of Jupiter, it is not clear whether the same considerations apply. The *Cassini* spacecraft in orbit round Saturn was not equipped with a Gravity Science instrument as sophisticated as that on *Juno*. At the moment, we must suspend final judgement pending results from future spacecraft missions.

Juno also raised new questions about the nature of the core within Jupiter. Theoretical models had predicted either a small,

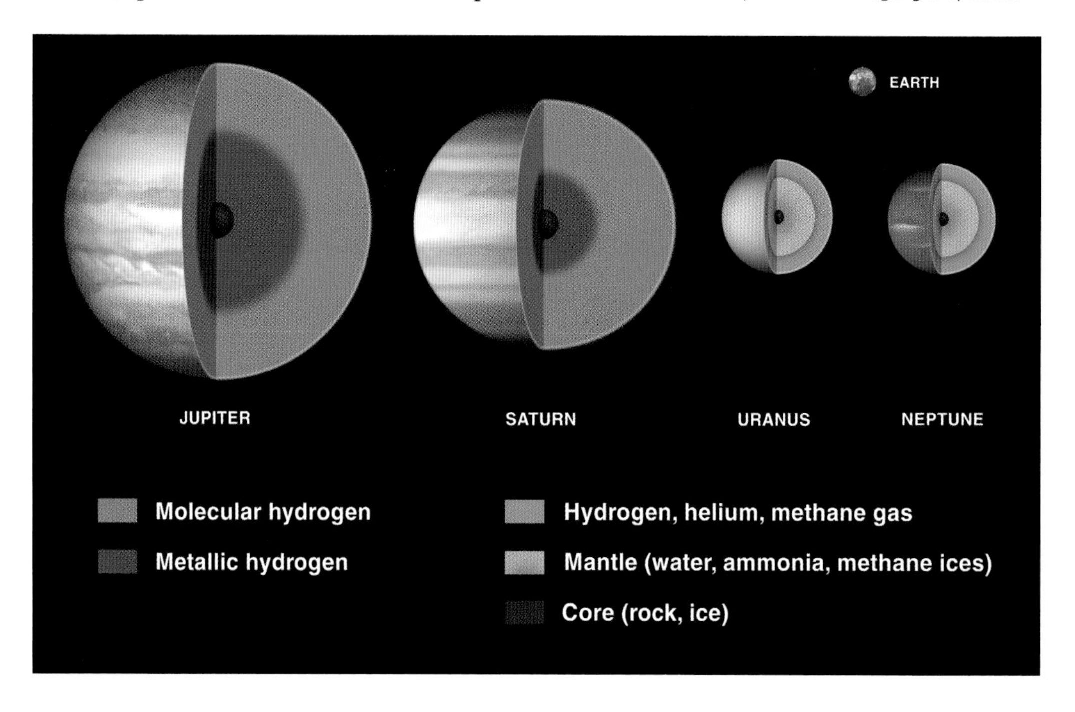

Models of the interiors of the gas giant planets.

Juno spacecraft image of Jupiter, taken by the NASA spacecraft on the outbound leg of its twelfth close fly-by of the planet, on 1 April 2018.

rocky core or no core at all, but *Juno* found evidence of a large, fuzzy core, which may be partially dissolved or perhaps disrupted by deep motions and zonal winds. Again, it is not clear whether Saturn will prove to be similar.

Glowing Worlds

On Earth, the driving energy source of the circulation and weather is the Sun. Though solar energy plays some role even at the distances of the giant planets, and in particular the tilt of Saturn away from or towards the Sun and the shadow effects of the rings give rise to seasons, it is not the whole story. As was discovered by infrared astronomers beginning in the late 1960s, Jupiter and Saturn emit more thermal energy than they receive from the Sun. (Later it was found that Neptune does so as well, but not Uranus, curiously.) Here, the nineteenth-century astronomers like Proctor scored a palpable hit.

Jupiter radiates twice as much energy as it receives, Saturn about 2½ times. The excess thermal energy emitted by Jupiter can be explained satisfactorily by slow gravitational contraction; essentially, the planet gives off heat as the matter making it up becomes more and more compressed. Saturn, however, does not appear to be massive enough for this process to work, and other factors must be involved (which may well be the case with Jupiter also). Helium 'rain' deep inside Saturn probably contributes: as helium droplets descend through the lower-density hydrogen, the friction produces heat, leaving helium-depleted outer layers (including the upper atmosphere; refer to the table on p. 78) and a helium-enriched shell around the core. Even more exotically, it has been suggested that rainfalls of diamonds may occur in the interior of Saturn, and in the other giant planets as well.

The Formation of Saturn: Rings to Worlds

The structure of Saturn – from its deep interior to its atmosphere – is of tremendous interest, but from the point of view of the solar system as a whole, all this is merely detail. In the big picture, Saturn's significance lies in its mass – after the Sun and Jupiter, it is the most massive body in the solar system. In Earth masses, the Sun, of course, ranks first (at 332,946 Earth masses), followed by Jupiter (at 318) and then by Saturn (at 95).

In 1833 the English astronomer George Biddell Airy wrote, 'Next to the elements of the planetary orbits, the most important numerical value for the explanation and prediction of the phenomena of the Solar System, is the mass of Jupiter.'[12] He might have continued that, after the mass of Jupiter, the next most important numerical value is the mass of Saturn. These two planets together contain more mass than all the rest of the bodies of the solar system (excepting the Sun) combined, and have had a dominant effect on the way the solar system has evolved from the very first.

The two planets exert marked perturbations on each other, which produce irregularities in their motions that were already felt by Kepler as early as 1625. He found that the tables of Jupiter made the mean motion (the rate of motion if it moved in a circular orbit) too slow, and that of Saturn too rapid. In other words, Jupiter seemed to be speeding up, and Saturn slowing down. Edmond Halley (1656–1742) verified this result, and attempted to explain it on the basis of their mutual gravitational attractions. Later astronomers showed that the acceleration of Jupiter necessarily would be offset by a retardation of Saturn, but also wrestled with a further, embarrassing result: the solar system ought, in the course of ages, to lose its two most prominent members, with Jupiter falling into the Sun and Saturn being driven off into the depths of space.

The situation was finally clarified by Laplace in 1785. He began with the well-known fact that Jupiter's period of revolution around the Sun, 11.86 years, and Saturn's, 29.57 years, are very nearly in the ratio 5:2. In other words, Saturn completes almost exactly two revolutions for every five completed by Jupiter. As with Mimas' orbital motion relative to particles in the Cassini Division, this represents a near-resonance situation. The 5:2 ratio means that after twenty years (more precisely, after 19.86 years), Jupiter and Saturn line up together (are in conjunction) in the same position relative to the Sun. However, each conjunction is displaced 120° towards the west from the last one, so that after three successive conjunctions, or sixty years, they come together again in the same part of the sky. (Remember, it had been this relationship that Kepler had been sketching on the blackboard before his students when he had the inspiration that led him to the Platonic solids as his explanation of the structure of the solar system, as described in Chapter One.)

False-colour composite image combining visual and infrared data shows an auroral glow at the south pole of Saturn (greenish) and thermal emission from the interior of Saturn (reddish).

If the 5:2 resonance were exact, the conjunctions of Jupiter and Saturn would repeat in exactly the same place every sixty years. Then there would be a pull in the same direction each time, and after many revolutions there would be a transfer of momentum and a shift of orbits, just as occurs between Mimas and ring particles in the Cassini Division. However, the Jupiter/Saturn orbital resonance is not exactly 5:2 but much more nearly in the ratio of 72:29. This means that though the conjunctions take place almost 120° apart, each set of three is shifted 8.37° from the last, and rather than a resonance, we have a near-resonance. But near-resonance relationships have no dynamic significance. After each cycle, the relative position of the bodies shifts, and so over astronomically short timescales, their relative positions are as random as those of bodies that are nowhere near resonance. In the case of Jupiter and Saturn, as Laplace showed, there was an apparent retardation of Saturn and an apparent acceleration of Jupiter that reached a maximum in 1560; then they began to decrease towards the values of their mean motions (their average rate of motion if they moved in circular orbits), which they attained in 1790. Afterwards, the motion of Saturn accelerated, while that of Jupiter was retarded. In other words, for 450 years Saturn's orbit is gradually enlarging and Jupiter's shrinking, and then, during the next 450 years, the situation is reversed, with Saturn's orbit shrinking and Jupiter's enlarging. The effects cancel out, and there is no net change over the whole period of nine hundred years. This result was reassuring. Jupiter and Saturn remain in the solar system.

Indeed, one of the major preoccupations of Laplace and other astronomers of his time was to establish that the solar system is 'stable' – that, though the planets experience periodic changes owing to gravitational perturbations, the changes are confined within narrow limits, so that the orbits of Jupiter and Saturn, within these limits, were supposed to have remained more or less the same since the solar system formed.

Another assumption was that if other planetary systems belonging to distant stars were ever found, they would closely resemble our own solar system, with giant gaseous planets situated far from their stars, moving in circular orbits like that of Jupiter, and with the smaller, rocky planets like the Earth lying closer in.

This assumption was only natural, given that our solar system was the only one we knew. Statisticians warn that such cases can lead to 'misinterpretation of incomplete and unrepresentative data'.[13] With only one Earth – and one solar system – to judge from, there was simply no way of knowing whether those cases were representative or not – an example of what physicist Frank Wilczek has called the problem of 'projection', ascribing 'universal significance to what is actually a very limited slice of the world'.[14]

Over the last two decades, since 1995 when the first extrasolar planet (51 Pegasi b) was discovered, our slice of the universe has indeed become much bigger. Using one of two methods – the first, the Doppler method of looking for shifted lines in the spectra of stars, or the second, the transit method of watching stars dim as planets in our line of sight move across them – thousands of exoplanet systems are now known. The vast majority of these have been found by NASA's *Kepler* spacecraft, launched in March 2009.[15] Kepler's only instrument is a photometer used continually to monitor the brightness of approximately 150,000 stars in a single, fixed field of view. Planets are detected whenever they pass in front of their stars, and produce a periodic dimming in brightness. At the time of writing (April 2019) 4,000 exoplanets have been discovered, with many of them belonging to multiple planet systems, like our own solar system.[16] Far from being unique, the Earth is but one of countless planets, and even conservative estimates suggest there may be many times more Earthlike planets in the galaxy than people living on the Earth.

The sheer ubiquity of planets around other stars proves that, in contrast to early twentieth-century ideas of the formation of planets which invoked rare encounters between stars, planet formation must be commonplace, and at least in broad terms, there is now a growing consensus about how this happens. The solar system formed from material that, in contrast to the pristine hydrogen and helium (and a smidgeon of lithium) that had been formed in the Big Bang, was processed over billions of years of stellar evolution in the galaxy, until it consisted of the mix of elements we refer to as the 'solar' abundance. This material was present as gas and dust in a large interstellar molecular cloud, like the 'dark nebulae' that appear in projection against the Milky Way. Some 4.6 billion years ago – which is the age of the solar system – one of these molecular clouds underwent collapse, either through gravitational instability or through the triggering shockwave of a supernova explosion. (The timescales of collapse differ in these two scenarios: gravitational collapse would take about 10 million years, while triggered collapse is more rapid, and would take only about a million years. At present, the supernova-triggered collapse scenario seems somewhat more probable. It also explains why the solar system is 'metals' rich.)

As the material of the cloud began falling inwards, its gravitational energy was converted into heat. The centre eventually became hot enough to trigger thermonuclear reactions in the core. The resulting radiation pressure was able to resist further gravitational collapse, so that a star – our Sun – was born. The Sun has remained in fairly steady balance ever since. Meanwhile, the angular momentum of the collapsing cloud gave rise to a spinning circumstellar disc of gas and dust.

The material in the disc is the raw material of planet formation. Apparently, with the cooling of the disc, heavier elements, such as silicates and iron, condensed out, settled into the plane of the disc, and began to collide with one another and stick together to form larger bodies (known as planetesimals). Some of the planetesimals lying at a distance from the star where there are significant amounts of water ice became large enough and massive enough to collect large amounts of gas before the star heated up to the point where it began to blow the rest of the gas in the disc away. These gas-gorging bodies became giant planets, like Jupiter and Saturn. This had to happen quite quickly, since it appears to take less than 10 million years for most stars, and in many cases, less than a few million, for gas loss to occur in the discs. Indeed, gap-clearing in these discs – interpreted as evidence of planet formation – has been directly observed in some young planetary systems, of which one has been estimated as only about a million years old.[17]

In addition to giant gas-rich planets like Jupiter and Saturn, small, rockier lumps, lying closer to the star, where water ice is unstable, are unable to collect much gas and, therefore, become the Earthlike terrestrial planets – mere naked, rocky cores without significant gaseous envelopes. The arrangement in our own solar system, with the small rocky planets lying close to the Sun and the gaseous giant planets lying farther away, fits logically with this scheme.

Because the Doppler method works well in disclosing the existence of giant planets close to their stars, these were the first to be discovered. However, this was merely a result of sampling bias; analysis of the hundreds of exoplanets now discovered by the *Kepler* spacecraft shows that only about 5 per cent have gas giants like Jupiter and Saturn. Also, in contrast to what is found in our solar system, where Jupiter and Saturn move in quite circular orbits, many of the giant planets discovered by *Kepler* follow highly eccentric paths, and in a few per cent of cases lie even closer to their stars than Mercury does to the Sun. These, called 'hot Jupiters', made up a disproportionate number of the early discoveries, simply because they were the most easily detectable with the Doppler shift method. Though the technology needed to detect them was available to exoplanet searchers as far back as the 1960s, they were overlooked, simply because no one expected to find Jupiter- and Saturn-mass objects so close to their stars.

The bizarre planets though did begin to turn up, and they seemed rather like something out of *Alice's Adventures in Wonderland*. As Alice said to the Queen, 'There's no use trying; one can't believe *impossible* things.' These gas giant worlds had somehow managed to grow to such enormous size despite the fact that there was no ice close enough to their stars.

Since, however, there is no way for Jupiter- or Saturn-sized planets to form without the condensation of water ice, astronomers were presented with a puzzle. The solution came with the realization that these giant planets could not have formed at their present distances so close to their stars, but instead must have formed much farther out – beyond the 'ice-line'. In other words, giant planets do not necessarily remain in the positions in which they formed, but are able to migrate. In the case of the 'hot Jupiters', they have migrated inwards towards their stars.

The way this is thought to happen involves, once again, orbital resonances. We now know from the study of exoplanet systems that

Concentric bright rings separated by gaps in the circumstellar disc of HL Tauri, a star which is no more than a million years old but whose disc already appears to be forming planets.

the formation of giant planets like our own Jupiter and Saturn is quite rare, but when it does occur, such giant planets interact gravitationally with the circumstellar disc and begin to migrate inwards. Computer models have shown that they typically end up at an orbital distance from their star similar to the Earth's from the Sun.

This did not, however, happen in our own solar system, which appears to represent a rather unusual case. It was similar in the first act, but different in the second. Jupiter, the first planet to form, did indeed begin to interact gravitationally with the circumstellar disc, and to migrate inwards towards the infant Sun. Our solar system would then have looked well on its way to becoming a typical 'hot Jupiter' system, and we would not exist. Its inward migration was arrested by the formation of the second planet to form, Saturn, which also began migrating inwards, but because of Saturn's lesser mass, did so more rapidly. As Saturn careened inwards and the two planets came closer together, they arrived at an orbital resonance – not the nearly 5:2 resonance of today but an exact 1:2 resonance, like that between Mimas and particles in the Cassini Division, between Io and Europa in the Jupiter system, or between Enceladus and Dione in the Saturn system. Such a resonance can either stabilize (sometimes only temporarily) or destabilize the orbits of the bodies involved. In the case of the particles in the Cassini Division, they are destabilized, as we have already seen. In the case of the satellites of Jupiter, in addition to Io and Europa, a third body, Ganymede, is also involved. These three are involved in a three-body resonance (known as a Laplace resonance, after the discoverer), where the orbital ratio is 1:2:4. The perturbations of Io and Europa on one another, which by themselves would be destabilizing, are exactly compensated for by the perturbations by Ganymede. Thus the resonance is stable, and indeed the three satellites have been observed to maintain their relations exactly, without the slightest deviation, in the thousands of revolutions they have made since they were discovered by Galileo in 1610. In the case of Enceladus and Dione, the 1:2 resonance is temporary, and eventually these moons will evolve into different orbital relations.

Returning to Jupiter and Saturn in the early history of the solar system, as soon as they arrived at the 2:1 resonance, their mutual gravitational disturbances on each other and on the remaining disc material were greatly enhanced. This led to the formation of a gap in the disc, which separated it into two parts. The inner disc was dominated by Jupiter, the outer by Saturn. Because of its greater mass, Jupiter exerted more of a pull on the inner disc than Saturn did on the outer, and seems to have migrated inwards as far as the present orbit of Mars, scattering debris (planetesimals) as it went and tossing some into the inner solar system, where they would later seed the still-unformed inner planets (including the future Earth) with water and other volatiles. Saturn continued to migrate, more slowly, through the outer disc. The result of these migrations was to decimate the planetesimal population and to exchange enough angular momentum to cause a reversal of direction, known as the 'Great Tack', after which

Jupiter and Saturn began to migrate steadily outwards rather than inwards. There they began to interact gravitationally with the two more recently formed giant planets, Uranus and Neptune (and probably one or more gas giant planets that were ejected from the solar system altogether). The migrations of these giant planets produced resonances 'sweeping' through the system, causing rapid rearrangements of planetary orbits and producing a large population of irregular bodies in eccentric orbits, some of which were later captured as satellites by Jupiter and Saturn. They also disturbed some of the planetesimals in the disc and flung them across the solar system – not as a Late Heavy Bombardment, but a more or less steady onslaught beginning about 4.5 billion years ago and petering out about 3.8 billion years ago when such classic features as the Mare Imbrium basin on the Moon and the Hellas basin on Mars formed.

Though Jupiter is, after the Sun, the dominant mass in the solar system, Saturn's role was also critical, for it was Saturn that pulled Jupiter back and steered the solar system in a different direction from most of the known solar systems with gas giant planets. This was to have enormous consequences – for life on Earth, and ultimately for ourselves. According to recent work by Konstantin Batygin (now at Caltech) and Gregory Laughlin (Princeton), at the point of Jupiter's farthest inwards migration, the planetesimal population would have been depleted and ground down to boulders, pebbles and sand. Any primordial inner planets such as 'super-Earths' (planets several times more massive than the Earth), which are proving commonplace around other stars, would have been pushed into death spirals ending in the Sun. All that would have survived would have been a sparse and narrow ring of rocky debris. From that rocky debris, Mercury, Venus, Earth and Mars would form.[18] There is a freakish aspect to our solar system: terrestrial planets close to the Sun, as those in our solar system are, with relatively low masses and thin atmospheres, are apparently rather rare, if not quite unique.

Scriven Bolton (1883–1929), an English amateur astronomer and pioneering space artist from Leeds, made this evocative rendering with a 25-cm reflector on 18 December 1910. It served as the frontispiece to George F. Chambers's book *Astronomy* (1910).

The existence of the circumstellar disc, the formation of the giant planets and the ensuing gaps produced by resonances remind us – though scaled up grandly – of Saturn and its ring system. The analogy is a deep one, and there is more than a little truth in the dictum of Ormsby Macknight Mitchel, 'Saturn's rings were left unfinished to show us how the world was made.'[19]

Saturn is, arguably, the most beautiful object in our skies. We may well owe to it our very existence.

FOUR

Aesthetic Rubble (the Rings)

By the end of the nineteenth century, the complexity of the rings was beginning to be realized. As Maxwell had shown mathematically, and Keeler observationally, they consisted of billions of tiny satellites, moving in Keplerian orbits around Saturn, and being perturbed by the planet's inner moons (Mimas, especially). Kirkwood's resonance theory, which had accounted so well for the Cassini Division, was extended to explain other ring divisions. Encke's division, for instance, which appeared to be more complicated than just a single stripe and apparently consisted of a double minimum near the middle of the ring, was identified with a 2:5 Enceladus resonance and a 3:5 Mimas resonance, respectively, by Kirkwood himself.

The effort to extend Kirkwood's theory to minor ring divisions was pushed farthest by the American astronomer Percival Lowell (1855–1916), today best remembered for his theory of intelligent life on Mars and his mathematical calculations for the trans-Neptunian planet 'X'. In 1915, when the rings were wide open and Saturn far north of the celestial equator, he and his assistant Earl C. Slipher (1883–1964) trained the 61-cm Clark refractor at his observatory at Flagstaff, Arizona, on the rings and noted numerous fine divisions. The B ring in particular appeared, in Lowell's words, 'conspicuously striped amidst its shading', with the 'dark curving lines of its plaided pattern proving so definite as to permit of measurement'.[1] He found that though most of the stripes were near resonance positions with Mimas, they failed to line up exactly, even when the oblateness of the planet's figure was taken into account, and concluded that the only possible explanation was that the interior of the planet was not homogeneous but instead consisted of a series of shells having differential rotations. The interior was, in his phrase, rotating 'like an onion in partitive motion'.[2] His explanation was at least plausible, but it has now been completely ruled out by spacecraft observations.

Lowell used a powerful telescope in excellent atmospheric conditions, and seems to have had a real penchant for making out fine linear (or circumlinear) details. (Think of the canals of Mars!) However, his perspicuity in the detection of fine ring details was emulated by others, such as the French astronomer Bernard Lyot

Percival Lowell's sketch of Saturn from 1915, showing divisions in the rings as measured by himself and his assistant, Earl C. Slipher.

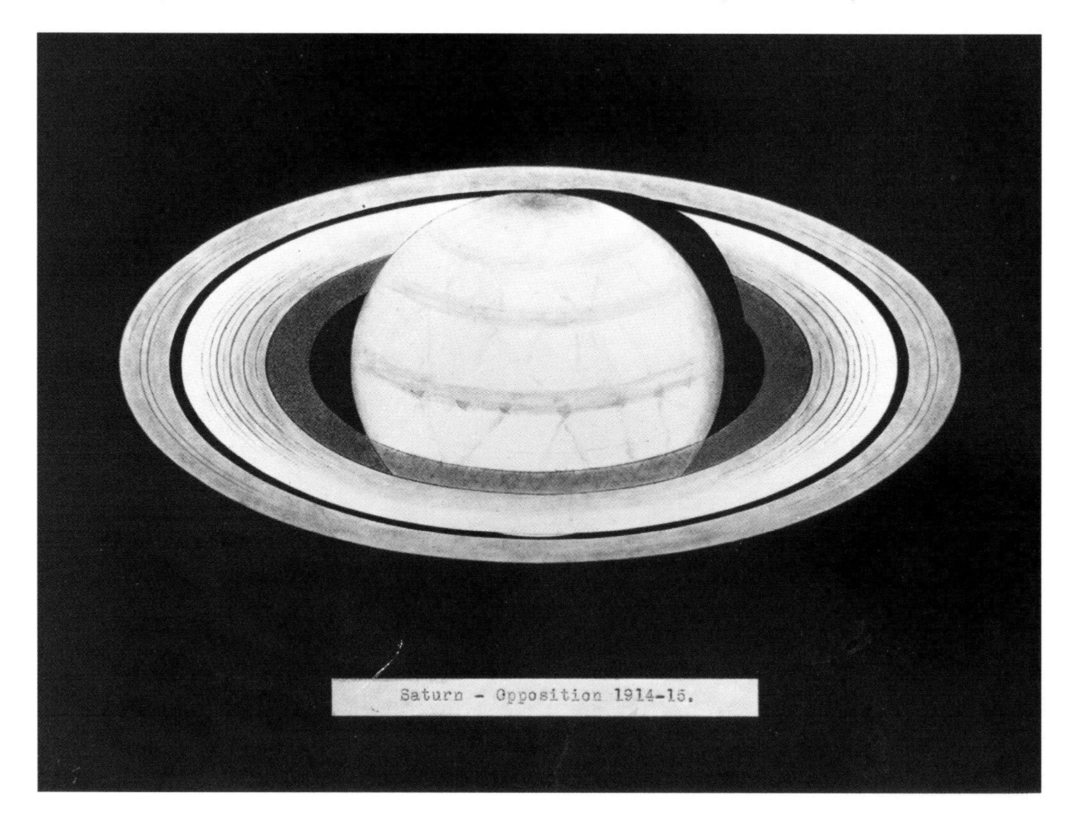

Bernard Lyot's 1943 drawing showing ring divisions as observed with the 61-cm refractor of the Pic du Midi Observatory.

(1897–1952), whose beautiful drawing from 1943, made with the 61-cm refractor of the Pic du Midi Observatory, was long regarded as the standard.[3]

But agreement about such features was far from unanimous. To some extent, the personality of the observer entered in as much as the instrument used, while detail visible under certain circumstances might be hidden from view at other times and places. Thus the famed Dutch American astronomer Gerard P. Kuiper (1906–1973), who, on a 'nearly perfect night', used 1,175× on the 5.08-m reflector at Palomar in California in 1954, reported finding only one genuine gap in the ring system – the Cassini Division. Encke's 'division' was a mere ripple in the position where the A ring changes intensity abruptly; there were also three ripples in the B ring, but he agreed with Barnard against Dawes that there was no gap between the B ring and the C ring.[4]

Still, in spite of Kuiper's authority and use of the largest telescope in the world at the time, the question can hardly be said to have been settled. Lyot's colleague, the French astronomer Audouin Dollfus, with the 2.1-m reflector at McDonald Observatory in Texas, reported seeing the rings in 1957, much as Lyot had depicted them. He also noted, as Lowell had done, that the divisions did not seem to agree exactly with the expected resonance positions, but instead of speculating about conditions in the interior of the planet, wondered whether this might be explained by assuming the existence of unknown inner satellites.

As is now known from spacecraft images, the rings are actually as grooved as a phonograph record. Even observers like Lyot and Dollfus grasped only a small part of the actual structure present, while Kuiper's negative report seems to have been affected, in large part, by the timing of his observation, which was made only days from the date of opposition – when Saturn is opposite the Sun and the shadow of the rings is hidden behind the ball. Conditions then are similar to those when the Moon is at full. Owing to what has been described as the 'opposition effect', the rings brighten dramatically at this time as the closer ring particles cover the shadows they cast on more distant ones.[5] To see minor divisions in Saturn's rings, it is best to observe several weeks before or after the opposition date, when the shadow of the globe on the rings is evident. These were the circumstances when (solar astronomer and very occasional Saturn observer) W. C. Livingston had a serendipitous view of Saturn under nearly perfect seeing conditions with the 2.54-m reflector at Mt Wilson in California. Though he did not recall the date, he said in 1975 that even Lyot's drawing did not come close to capturing the complex details, but 'the rings were fascinating, appearing as in an imaginative fine engraving. Each major ring was sharply bounded and contained numerous concentric internal subdivisions, resembling dark threads.'[6]

Not only are the fine divisions best seen far from opposition, but it is then, too, that the shadow of the globe on the rings is most conspicuous, providing an enhancement of the beauty of the scene with the added dimension of depth. As the poet Alfred, Lord Tennyson (1809–1892) put it in his 1842 poem 'The Palace of Art', 'While Saturn whirls, his steadfast shade/ Sleeps on his luminous ring.'

The closer to opposition Saturn is, the less conspicuous will the shadow be, while the maximum conspicuousness, in general, occurs when the angle between the line from the Earth to Saturn and from the Sun to Saturn is greatest (about 6°). Still another factor that affects the shadow geometry is the 26°.7 tilt of the rings to the plane of Saturn's orbit: at intermediate angles, the shadow will be more or less foreshortened and fail to reach all the way across the entire width of the rings, while only when the rings are nearly wide open does the shadow extend across the entire width of the rings.

Note that as the ring particles circulate in their orbits round Saturn and pass into the shadow of the globe, they undergo eclipse, and thus pass abruptly from sunlight into conditions of severe cold. From spacecraft observations, the rings' average temperature has been found to be some 85 kelvin (–188°C), but there are differences in temperature from one ring to the next, with the rings becoming cooler the further one travels from the planet, as they receive less heat from Saturn. (The temperature drops from 110 kelvin (–163°C) in the C ring to 70 kelvin (–203°C) in the B ring, though oddly, the A ring bucks the trend and is warmer than expected, at 90 kelvin –183°C). This is probably because bending waves produced by the gravitational pull of moons close to the ring edge raise material above the mean level of the ring material, where they are exposed to more sunlight.)

On entering the shadow of the globe, the ring particles are eclipsed for as much as two hours, which produces an additional 3 or 4 kelvin of cooling. The shadow effect also disturbs the ionized upper layer of Saturn's atmosphere, causing a reduction in the ionization of particles and allowing radio bursts caused by lightning to escape. Thus radio bursts surge along the shadow edge.

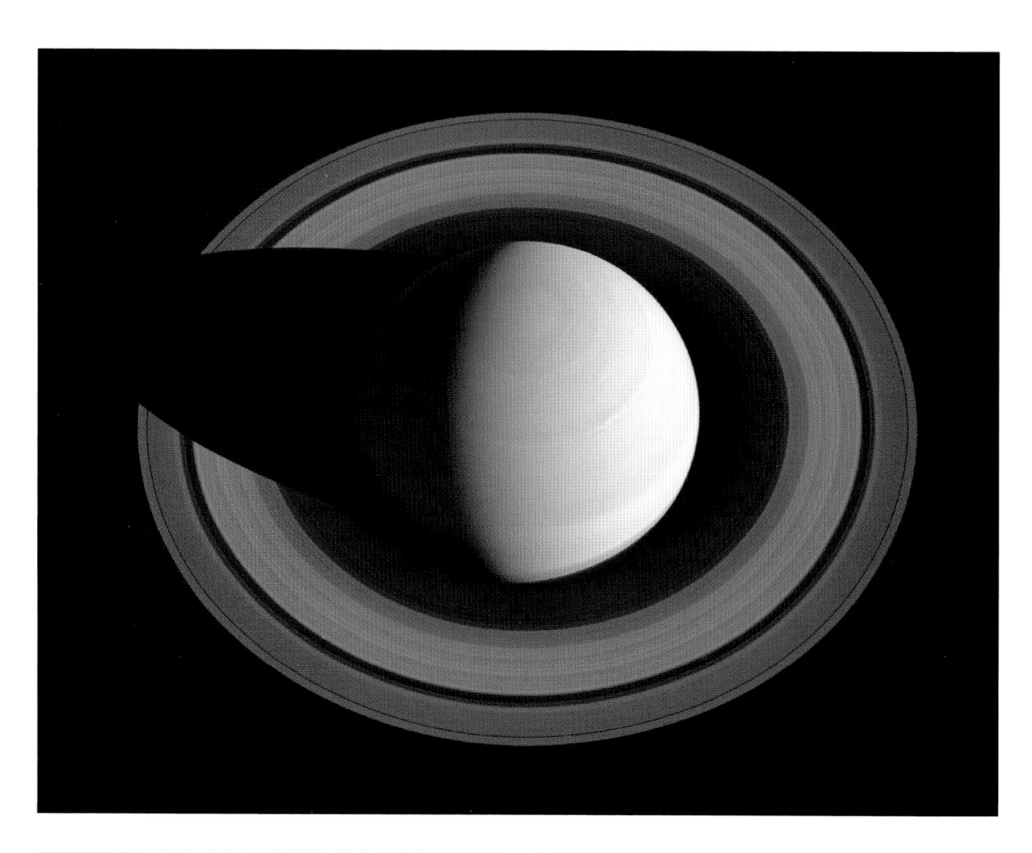

Occultations Reveal Ring Structure

In addition to visual observations using large telescopes in favourable seeing conditions, occultations by the rings have also been useful in clarifying their fine structure. Long ago, William Herschel suggested that the true nature of the Cassini Division could be determined directly if a star could be observed passing behind the rings and seen shining through the gap. He was, of course, quite correct, though curiously the first actual observation of such an event was not made until 9 February 1917, when two alert British amateur astronomers, Maurice Ainslie (1869–1951) at Blackheath and John Knight at Rye, achieved what Saturn historian A.F.O'D. Alexander hailed as 'one of the greatest triumphs ever achieved by observers of Saturn in Great Britain'.[7] Observing independently, they both happened to catch the rings' occultation of a seventh-magnitude star, and demonstrated not only the sparseness of material in the Cassini Division, but the translucency of Ring A through which the star continued to glimmer faintly. Indeed, using only a 23-cm reflector, Ainslie was even able to appreciate a strengthening of the starlight as it shone through two apparent gaps in the outer part of the A ring (the ones now known as the Encke and Keeler Gaps).

A similar observation of an occultation of an eighth-magnitude star was made by John E. Westfall (1938–2018), with the 51-cm refractor of the Chabot Observatory in Oakland, California, on 28 April 1957. (His interest in astronomy had been inspired at age seven when he saw a photograph of Saturn in an encyclopaedia.) Though only nineteen at the time of the occultation, he was already a seasoned observer, and was the only person to take advantage of the event and trace the star's visibility during its 3½-hour passage behind the rings:

> For about 10 min after entering the A ring, the star remained visible but fluctuated between full visibility to near-invisibility,

Cassini image taken on 10 October 2013, showing Saturn and rings from almost overhead. The shadow of the ball projects across the entire breadth of the ring system.

Cassini image taken on 25 April 2016. The planet was then approaching its northern hemisphere summer solstice, which took place in May 2017.

> then became invisible for about 10 min before recovering full brightness as it entered the Cassini Division. After first entering the B ring, the star remained faintly visible for another 10 min before fading completely, then stayed invisible except for a faint rise lasting about 1 min. The star was not seen again until it passed through the Cassini Division, and then was seen continuously during its outward passage through the A ring, although varying from near-invisibility to near-brightness.[8]

Subsequent observations have included occultations of the radio signals and of stars observed by the *Voyager* spacecraft in 1980–81, and the very favourable and widely observed occultation of the fifth-magnitude star 28 Sagittarii on 3 July 1989, in which the star passed behind not only Saturn's globe and all three major rings, but Titan (making this the first stellar occultation by that satellite ever observed), and observations from *Cassini* in orbit round the planet. The spacecraft observations, in particular, have led to remarkably detailed radial profiles of the optical thicknesses of the rings, which allow subtle changes in structure to be recognized over time.

Ghostly Rings

In addition to the three main rings (A, B and C), other rings, fugitive and uncertain, have been reported over the years. For instance, a 'luminous zone' beyond Ring A was reported by a leading French observer of the planets, Georges Fournier (1881–1954), with the 28-cm refractor of the Jarry-Desloges Observatory on Mont Revard in Savoy in September 1907, the year after the rings' edgewise presentation. There was an apparent confirmation of this outer dusky ring, referred to as Ring D (later as D'), by Émile Schaer (1862–1931) of the Geneva Observatory the following year. However, Barnard, at the same time, failed to see anything unusual with the Yerkes 1.02-m refractor.

A simulated image of Saturn's rings, using colour to represent information about ring particle sizes in different regions based on the measured effects of radio signals of 0.94-, 3.6- and 13-cm wavelengths sent through the rings.

An apparent faint extension outward from the A ring was also reported on the basis of long-exposure photographs taken during the edgewise presentation of 1966 by Walter A. Feibelman (1925–2004), while searching for faint satellites using the 76-cm Thaw refractor of the Allegheny Observatory in Pittsburgh. Then, too, there was an apparent, very faint inner ring, inside the crêpe ring and separated from it by a narrow gap, reported by the French astronomer Pierre Guérin on the basis of photographs taken in 1969. It was (somewhat confusingly) referred to as the D ring, while Feibelman's ring, not definitely identified with that seen by Fournier and Schaer, was referred to as the E ring.

Things were thus in a rather confusing state, and not until the spacecraft era did they get sorted out. During its 1979 fly-by, *Pioneer*

Approaching Saturn on 1 September 1979, *Pioneer 11* looked on the unlit northern face of the rings.

11 discovered a new ring (Ring F) and also confirmed Feibelman's E ring. From Earth-based observations during the edgewise ring presentation of 1979–80, astronomers Stephen Larson and William A. Baum (1924–2012) followed this ring all the way from the orbit of Mimas out to 8 Saturn radii (near the orbit of Rhea); the peak brightness was at the orbit of Enceladus, and they suggested – presciently – that Enceladus might be the source of the particles making up the ring. Finally the existence of the D ring was also confirmed.

The Rings in the Spacecraft Era

What has been described so far represents the results of some three centuries of surveillance of the rings, from the first recognition of what they were by Christiaan Huygens and G. D. Cassini using the most powerful research instruments of the seventeenth century. Significant progress was made. However, a definitive view of the rings awaited the spacecraft era.

The first spacecraft to Saturn was *Pioneer 11*, launched in April 1973 from Cape Kennedy (as Cape Canaveral was known between November 1963 and May 1973). After reaching Jupiter in December 1974, it was given a gravitational assist towards Saturn, where it arrived in August 1979. Though *Pioneer 11*'s images were not much better than the best Earth-based ones, the spacecraft proved that it was possible to reach as far as Saturn with a functioning spacecraft – something that only a few years before would have seemed the stuff of science fiction. It also provided views such as had been possible only briefly, and under exceptional circumstances, from Earth, notably, views of the dark side of the rings as seen from beyond the planet.

Though no trace of either the D or E rings reported by Earth-based observers by *Pioneer 11* (that awaited later spacecraft) were found, its dark-side image did reveal a narrow ring, less than 800 km across and located just 4,000 km outside the A ring. It had never been seen before. This was the F ring, which proved to be quite unlike any other ring known at the time: narrow, eccentric and inclined to the equatorial plane. The closest analogues were the set of narrow rings that had been discovered in Earth-based observations of a stellar occultation by Uranus just two years earlier.

The same *Pioneer* image that showed the F ring also showed a 200-km satellite, nicknamed 'Pioneer Rock'. It might be Janus, the tenth satellite of Saturn discovered by Dollfus at the Pic du Midi Observatory during the 1966 ring passage, whose orbital period – and hence position – had remained somewhat uncertain. Another possibility was that it was the largest of a swarm of small bodies suspected by ring theorists to lie between Mimas and the A ring. Clearly further observations were needed.

At the time of *Pioneer 11*'s arrival at Saturn, two much more sophisticated spacecraft, the *Voyagers*, were already under way. They were to make an epic 'Grand Tour' of the outer solar system by means of a series of gravity assists, being flung, slingshot-wise, from one planet to the next. Both *Voyagers* were to explore Jupiter and Saturn; *Voyager 2* alone would head on from Saturn to Uranus and Neptune. The principle of a gravity assist is to send a spacecraft on a trajectory where it just misses one planet, is sped up by the planet's gravitational pull, then heads on at higher speed to the next. The increase in speed of the spacecraft is not free, but purchased by theft of a tiny amount of gravitational energy from

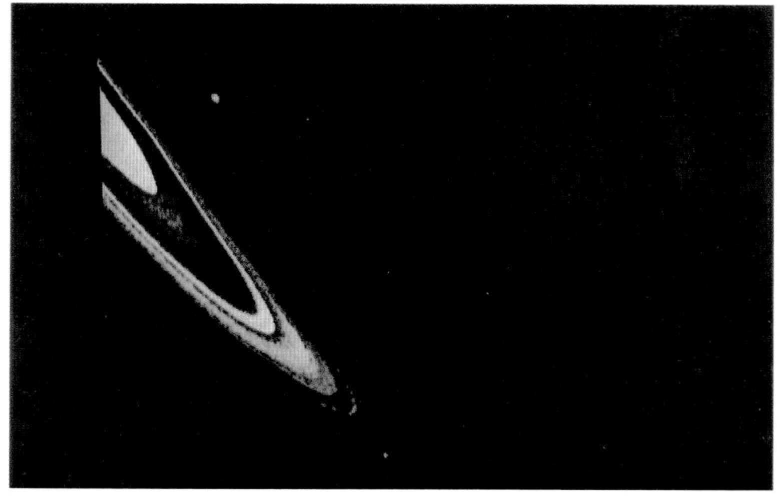

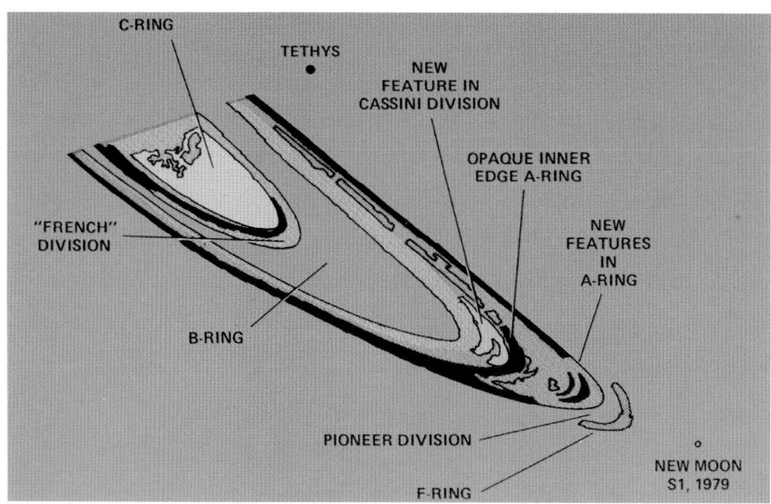

Pioneer 11's discovery image showing the F ring and the moon dubbed 'Pioneer rock'.

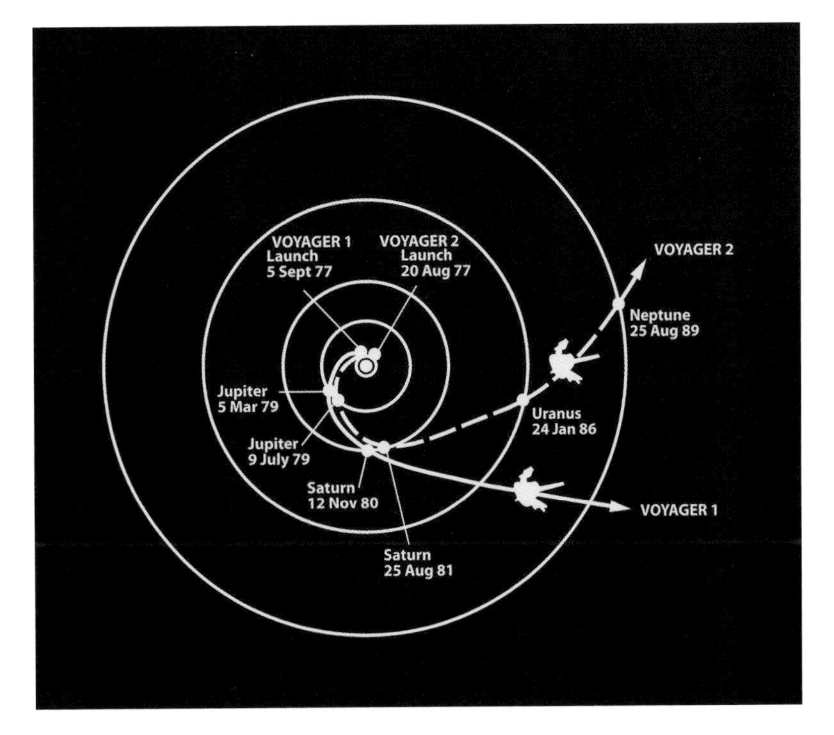

The trajectories of the two *Voyager* spacecraft through the outer solar system on their 'Grand Tours', swung from one giant planet to the next by means of gravity assists.

the orbital motion of the giant planet, and serves to greatly reduce the fuel requirements and transit times for the journey. This clever trick had first been demonstrated in the case of the *Mariner* 10 fly-by of Mercury in 1974, where Venus provided the gravity assist. For *Voyager*, the gravity assists from Jupiter to Saturn took advantage not only of the 'triple conjunction' of those two planets during 1980–81, but of a further fortuitous alignment of Uranus and Neptune in the same general direction. The next similarly favourable alignment will not, in fact, occur for another 170 years. (Pluto was also in a favourable position, but observing Saturn's satellite Titan at close range was a higher priority at the time; this meant sending *Voyager* 1 on a course that was bent far southwards out of the plane of the solar system, and nowhere near Pluto.)

Voyager 1 was launched from Cape Canaveral on 5 September 1977, received a gravity assist from Jupiter at its March 1979

Bradford A. Smith, *Voyager* Imaging Team leader, with Carolyn Porco, future *Cassini* Imaging Team leader, at Caltech's Jet Propulsion Laboratory during the *Voyager 2* Neptune encounter, August 1989.

encounter, and then swung by Saturn in November 1980. *Voyager 2*, though launched sixteen days earlier, on 20 August 1977, traversed a longer trajectory, taking it past Jupiter in July 1979, Saturn in August 1981, and then on to Uranus in January 1986 and Neptune in August 1989. Both performed heroically, and during their Saturn encounters, revealed the ball, the rings and the satellites of Saturn in unprecedented detail.

Though a great deal was learned about the atmosphere and interior of the planet, as summarized in Chapter Three, the most spectacular findings concerned the rings and the satellites, for which the *Voyager* Imaging Team leader Bradford A. Smith (1931–2018) deserves much of the credit. He was one of the few who, in the words of his *Voyager* colleague and later *Cassini* Imaging Team leader Carolyn Porco,

> had the foresight to recognize that the satellites and, later, the rings of the outer planets would be as fascinating as the planets themselves, and the need for a high-resolution imaging capability to address both. These realisations led him to insist

> on a change in the optics of the *Voyager* cameras and to hand pick, in addition to the original NASA-selected imaging team additional scientists with expertise in geology and planetary rings, and also scientists who were directly involved in ground-based studies of the bodies that *Voyager* would visit. Many of us involved in *Voyager* and subsequently chosen for the *Cassini* mission were among those added to the *Voyager* team by Brad.[9]

Even before the *Voyagers* had arrived at Saturn, the theory of the rings' structure had moved decisively beyond the pioneering work of Daniel Kirkwood. Scott Tremaine, who had worked on the structure of galaxies for his doctoral dissertation, finished in 1975 at Princeton University, had then taken up a postdoctoral fellowship at Caltech under Peter Goldreich. Tremaine wanted to do something completely different from galaxies, and asked Goldreich if he had any suggestions. Goldreich suggested that he work on planetary rings.

Tremaine began by looking closely at the Cassini Division, in an effort to understand better the gap here. Kirkwood's resonance theory had provided a qualitative explanation. However, physicists are generally not satisfied with qualitative explanations, and like to calculate such things exactly. At the time, computer models, using simplifying assumptions, had enjoyed some success, but they consistently predicted narrower gaps at resonance positions than those actually observed. Thus the challenge was to figure out how to widen the Cassini Division from a narrow gap about 30 km wide near the Mimas resonance position, which was what the available theory predicted, to the observed value of some 4,800 km.[10]

Tremaine and Goldreich proposed a solution in which the structure of Saturn's rings was regarded as analogous to that of spiral galaxies. In galaxies, spiral arms consist of gas, dust and stars, which during the course of a galaxy's lifetime rotate through perhaps one hundred revolutions. All these turns might be expected to cause the spiral to tighten towards the centre, but that is not what one sees: instead, there are only one or two accumulated turns. In other words, the galaxy appears to act like a rigid pinwheel. The theoretical explanation for this was put forward by C. C. Lin of MIT and Frank Shu of Berkeley in 1964. They argued that the spiral arms are actually only waves of more densely packed matter – spiral density waves, propagated by their own gravitational attraction for other particles rather than by particle collisions.[11] These spiral density waves, also known as quasi-static density waves or 'solid sound waves', pass through the galaxy like the waves of the ocean under a boat, and in the case of the Cassini Division, Tremaine and Goldreich realized, Mimas was setting up exactly the same kind of density wave.

At the 2:1 resonance position with Mimas, the particles in the rings are stretched into a bi-lobed oval. As particles swing into a lobe, temporary concentrations form, gravitationally disturbing nearby particles and leading to the production of outwardly spiralling density waves. In this way, angular momentum is transferred from Mimas to resonant ring particles, pushing them out. On the basis of this theory, Tremaine and Goldreich carried out a quantitative calculation that, at least roughly, accounted for the size and other properties of the Cassini Division.[12]

Things seemed to be falling into place, and even the fact that Saturn alone of all the planets had rings seemed to have an explanation. Mimas was the only major satellite in the solar system with a strong resonance inside the Roche limit. None of the other planets – Jupiter, Uranus or Neptune – had anything like it, and none of them at the time were known to have rings.[13]

As so often in science, the conclusion was premature, and was overthrown when astronomers discovered the suite of narrow rings, typically on the order of a few kilometres across, around Uranus at a stellar occultation in March 1977.[14]

Saturn's ring system – splendid beyond compare though it is – now had at least one rival; subsequent discoveries of rings

Diagram of a spiral density wave pattern, showing 'bunching up' of material, rather like cars during a traffic jam.

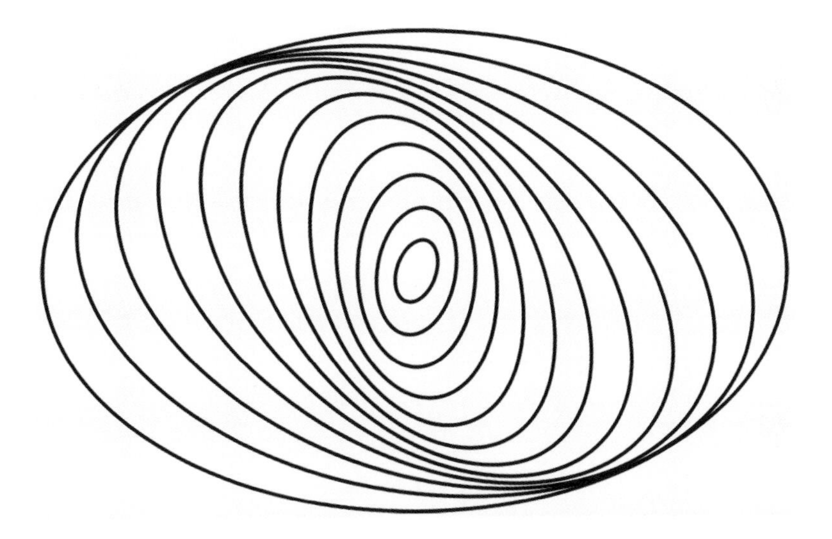

Cassini Division density wave.

around Jupiter and Neptune would show that planetary rings are not uncommon. (Indeed, at least one asteroid, 10199 Chariklo, which orbits the Sun between Saturn and Uranus, is now known to have a ring, and there are no doubt others.) Having begun their research by trying to explain why the Cassini Division was as broad as it is, Tremaine and Goldreich now faced a further challenge: explaining why Uranus has such narrow, sharp-edged, rings. In fact, the real problem with rings is to keep them from spreading, and this was emphatically so in the case of the slim Uranian rings. Left to themselves, the billions of ring particles ought to collide with one another and disperse into a wide homogeneous sheet. How, then, were the very sharp edges for the Uranian rings, as well as for Saturn's rings – say, at the outer edge of the A ring – to be accounted for?

Tremaine and Goldreich published their theory to explain this just as *Pioneer 11* was nearing Saturn. The same angular momentum transfer mechanism that they had invoked to explain how the Mimas resonance produces the Cassini Division would also apply, they realized, to a satellite much closer to a ring. But in order to confine particles in a narrow ring, rather than simply push them out, two small satellites were needed, one located outside and the other inside the ring. The small satellite located just outside a ring would travel at a slightly slower speed than the ring particles, thus exerting drag on particles bumped outwards through mutual collisions, slowing them down and causing them to drop back into lower orbits. The small inner satellite, on the other hand, travelling faster than the ring particles, would slightly accelerate any particles jostling inwards, forcing them back into higher orbits. In short, satellites on either side would effectively 'repel' wayward particles and gravitationally steer them back, through a process that came to be called 'shepherding', since, as Tremaine put it, 'the satellite was like a sheep dog going around a flock of sheep, and "barking" at it gravitationally to keep it all in line.'[15]

Like Grooves on a Gramophone Record

It would be left to *Voyager 1* to confirm this theory, and to provide a more definitive view of the rings' complicated structure.

By midsummer 1980, *Voyager*, still 100 million km out, was already sending images better than any that had ever been obtained from the Earth. By October, a month before the closest passage, the rings were being revealed as far more complex than anyone had imagined. Though a few additional divisions at resonance positions were expected, the reality far exceeded expectation, and was more reminiscent of Charles Tuttle's nineteenth-century report of 'a series of waves in the rings' than anything else. For instance, the Cassini Division had at least one interior ringlet, while the main rings were resolved into a series of concentric rings that soon were being likened to the grooves on a gramophone record. How could all this remarkable structure be explained? The numerous ringlets might be produced by resonances with small undiscovered satellites which, despite having small mass, would have disproportionate effects owing to their acting at short distances. This might explain much of the structure, but it was early days yet.

A *Voyager 2* image taken on 22 January 1986, during its Uranus fly-by. This view shows the inner rings, of which the brightest, at the top, is called epsilon. Moving inward, the next rings are called delta, gamma, eta, beta and alpha. The three rings to the lower left, at the limit of detection of the *Voyager* camera, are known as the 4, 5 and 6 rings.

The Mysterious 'Spokes'

As it drew ever nearer to Saturn, *Voyager* added a further mystery about the rings. Its images showed a series of shadowy radial markings extending like

Rings in colour: *Cassini* image taken on 22 August 2009.

A very high-resolution colour image showing a portion of the inner-central part of the B ring, between 98,600 km and 105,500 km from Saturn's centre. Here, these features have extremely sharp boundaries on even smaller scales than the camera can resolve, but closer to Saturn the structures become fuzzier and more rounded, less opaque and less colourful.

A maze of lines. In this *Cassini* image, the rings hang above the ball, while faintly through them, as through a diaphanous veil, the shadows are seen in projection against the upper cloud deck.

The particles making up the rings rotate from sunlight into the utter chill of the shadow of the globe.

Crisscrossing lines – rings and shadows intersect in the line of *Cassini*'s sight.

fingers across the B ring and behaving like the rigid spokes of a wheel rather than following the expected Keplerian shear of independent particles. They ought not to exist – and yet there they were.

The reason why they were not expected has to do with the fundamental fact of the ring particles' differential rotation, predicted by Maxwell and demonstrated by Keeler at the end of the nineteenth century. Particles at Ring B's inner edge travel in Keplerian orbits having periods of 7.9 hours; those at the outer edge have periods of 11.4 hours. This means that even if a radial feature developed – and *Voyager* images showed such features extending as much as 10,000 km in length – it ought to be immediately disrupted by Keplerian shear. An important clue to the unravelling of the mystery came when it was realized that these spokes appeared at radial distances where the particles moved in periods that were nearly 'kronosynchronous' – to coin a term – that is, they were co-rotating with Saturn, specifically with its magnetic field. It was also noted that they were most apt to appear at a specific magnetic longitude that was roughly aligned with Saturn's morning terminator. It seemed plausible that the spokes were produced when tiny particles in the rings acquired an electrostatic charge, were levitated out of the ring plane by Saturn's magnetic field and were then swept along by the magnetic field lines. The existence of the spokes was one of the most startling of the discoveries made by *Voyager* – and yet, surprisingly, it may have been anticipated.

In part inspired by the legendary nineteenth-century Harvard observers of Saturn, W. C. and G. P. Bond, Charles W. Tuttle and Sidney Coolidge, and encouraged by Harvard astronomer (and ring expert) Fred Franklin, an amateur named Stephen James O'Meara began a careful study of the planet with the 23-cm refractor at Harvard in the mid-1970s. (The reason he used this instrument rather than the famous 38-cm, Merz and Mahler refractor used by the nineteenth-century observers was that the latter was out of commission owing to a problem with turning the dome.) As he recalls,

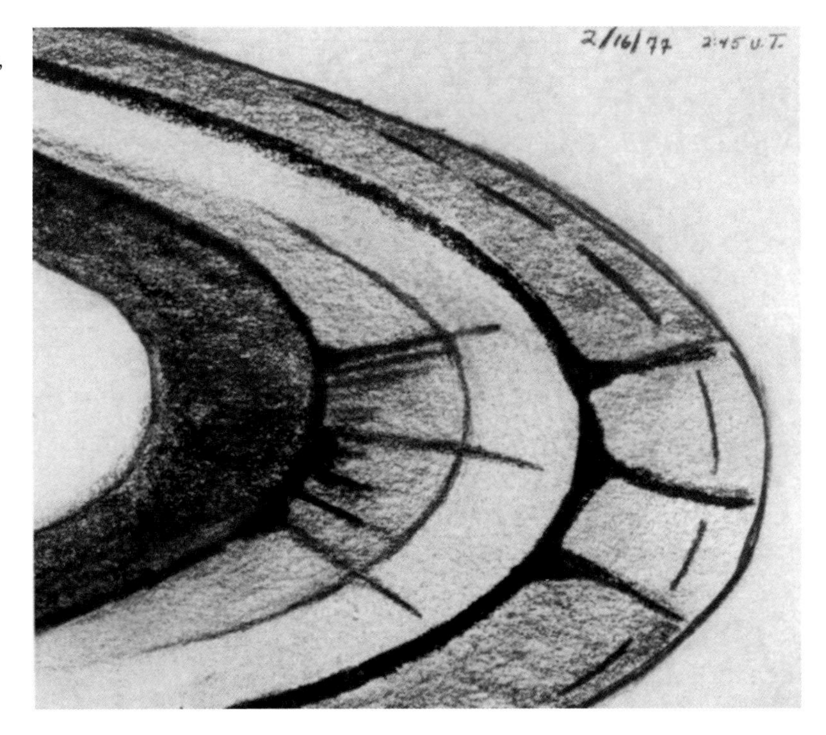

Radial structures in the inner B ring and the A ring, as sketched by Stephen James O'Meara in 1977. O'Meara called them 'spikes' at the time.

On 18 April 1976 I began a project to visually document dim (0.1 magnitude) azimuthal brightness variations in Saturn's A Ring for comparison with photometric measures. When the project ended successfully a month later, Saturn was heading toward conjunction with the Sun. Nevertheless, I found myself staring at the rings through the telescope, wondering if Saturn's B-ring also displayed any 0.1-magnitude brightness differences in azimuth. In time I found a different type of phenomenon – radial dusky bands which I called 'spikes'. When I informed Fred Franklin of this sighting, he was supportive but perplexed, and explained to me how, according to theory, the differential rotation periods of particles moving in Keplerian orbits would rapidly disrupt any radial features. Nevertheless, certain of their existence, I embarked on a long-term systematic (pre-spacecraft) study of them that ended in 1980, when Saturn and its near

Voyager 2 image taken on 4 August 1981, showing spokes in the B ring. They are just visible in the left ansa.

edge-on rings went into conjunction with the Sun – and just before *Voyager 1* arrived at Saturn and imaged Saturn's radial 'spokes'. My observations showed not only diurnal variations (appearing most intense on the morning ansa and being very sporadic and weak on the evening ansa), but they also varied in appearance and number from night to night and co-rotated with the planet. The view was also affected by ring tilt and illumination effects as seen from Earth.[16]

When the spokes first appeared on the screen in the Jet Propulsion Laboratory press room, O'Meara's colleague at

Sky & Telescope, J. Kelly Beatty, who was on hand covering the Saturn encounter for the magazine, had intimated that *Voyager* might have been forestalled – by an amateur astronomer no less. O'Meara had shown Beatty his drawings of the spokes beforehand, and had specifically suggested watching for radial features in Ring B. As journalist Mark Washburn recalls in his *Distant Encounters*, 'The news traveled quickly, and an hour later Brad Smith appeared in the press room, asking Beatty, "What's this I hear about someone seeing spokes?"'[17]

Though Smith remained sceptical, once attention had been called to them, the spokes were regularly recorded, including by amateurs.[18] Visual observers, including Clyde Tombaugh (1906–1996) with a homebuilt 41-cm reflector in his backyard in Las Cruces, saw them at times.[19] The first Earth-based CCD images of them were obtained with the 3.6-m Canada–France–Hawaii telescope (CFHT) atop Mauna Kea, Hawaii, in 1996, and thereafter by many amateur CCD imagers. Master CCD imager Donald C. Parker (1939–2015) of Coral Gables, Florida, often captured them, and they were regularly recorded by the Hubble Space Telescope from just before the ring-plane crossing in 1995 until October 1998.

After October 1998, they dropped from sight for several years. Presumably, they were still present, though perhaps rendered invisible because of the increasing inclination of the rings, the idea being that they might only be visible when the observer lay reasonably close to the ring plane. There was every confidence that they would be seen again when the *Cassini* spacecraft entered orbit round Saturn in 2004, but once again nature had other plans. (We will not give away the ending here but save it for later.)

Moon-tortured Rings

We now recall Tremaine and Goldreich's shepherd satellite theory. Just days before *Voyager* 1's closest approach to Saturn, a couple of irregularly shaped satellites, named Prometheus and Pandora, first showed up in some images. Prometheus is slightly larger at 102 km across, with Pandora's diameter at 84 km. They orbit just inside and just outside the fine line of the F ring, and proved to be 'shepherds' keeping the F-ring particles herded in a narrow 'flock', just as the Tremaine and Goldreich theory predicted. Meanwhile, another tiny moon, Atlas, was found orbiting just 1,000 km outside the A ring. It too was supposed to be a shepherd, in this case herding the particles in the outer part of that ring and producing its hitherto unexplained sharp outer edge.

On 12 November, *Voyager 1* made its closest approach, passing within just 124,000 km of Saturn's cloud tops. As it did so, it obtained remarkable new images of the F ring. *Voyager 1* had already confirmed a *Pioneer 11* result, finding two bright clumps in the ring, each some 1,000 km long, orbiting at the same rate as the ring particles, but the latest revelations were quite perplexing. As Brad Smith announced at a press conference, 'In this strange world of Saturn's rings, the bizarre becomes the commonplace, and that is what we saw on the F ring this morning.'[20] The ring appeared separated into strands, of which the two brightest appeared to be intertwined and braided.

By the following day, changes seemed to have taken place: the diffuse ring now appeared outside instead of inside the braided strands. Smith remarked that the kinks and braids appeared to defy the laws of celestial mechanics, and suggested, very tentatively, that perhaps some force other than gravity, such as an electrostatic force similar to that presumed to levitate particles to form the spokes, was involved. Just then, Goldreich, Tremaine and Stanley F. Dermott came to the rescue, showing that gravitational interactions between the co-orbitals alone might explain the strange structure. In previous research on the Uranian rings, Dermott had shown that as a satellite in the inner orbit caught up with one in the outer orbit, instead of colliding, the two would exchange orbital energy and, like trains switching tracks, the outer one would become the inner one and the inner one the outer. Prometheus and Pandora were the first cases where this was actually shown to happen – the pursued and pursuer switch places every four years. From the point of view of each one, the other traces a circular, horseshoe path every four years, which begins and ends with a close encounter.

Since these two co-orbital satellites are rather larger than necessary to confine the F ring, they produce waves; also, since the co-orbital satellites' gravitational attractions produce collisions among particles, with those involving larger, rock-sized particles taking longer to damp out than those involving smaller particles, the differential effects of gravity were presumed to segregate particles by size into the separate rings observed. One could sort

The F ring, as shown in this *Cassini* image, is corralled between Prometheus (inner orbit) and Pandora (outer orbit). Theories of ring formation suggest that multiple small satellites tend to form near the outer edge of the rings, and are susceptible to collisions. The dense cores are left behind as satellites like Prometheus and Pandora, and finer debris provides material such as that forming the F ring.

In this remarkable *Cassini* image from November 2010, shadows cast by 3.5-km vertical scale structures cast shadows across the outer edge of the B ring.

of see how these effects might produce structures like those in the F ring, though even Goldreich himself admitted that the kinks had him stumped. In fact, the dynamics of the F ring would prove to be far more complicated even than anyone imagined at the time, and were finally worked out only with new data from the subsequent *Cassini* mission.

Theorists were more successful in accounting for some of the other structures revealed in unprecedented detail by *Voyager*. The outer edge of the B ring, for instance, was found to correspond to the 2:1 resonance with Mimas, and further, the B ring's outer edge is neither circular nor elliptical but a two-lobed oval pointing directly towards Mimas and appearing to rotate with it about the planet. Each particle's path is a Keplerian ellipse; but because of Mimas' influence, at any given instant, any particle occupies a position along the two-lobed oval. The Mimas resonance gives rise to a spiral density wave – a series of particle-density fluctuations propagating outwards in the A ring. All this was in accordance with the Tremaine–Goldreich theory. However, because Mimas travels on an eccentric, inclined orbit (the orbital inclination is 1½°), the situation is somewhat more complicated than would be the case if it simply travelled in a circular orbit in the ring plane. Mimas is moving not only around the planet, but in and out, and up and down. Moreover, Saturn is an oblate globe rather than a point mass. Thus, though Mimas goes around Saturn once every 0.94242 days, it moves in and out (or up and down) in a slightly longer period of 0.94490 days. Because of this difference, the point of Saturn's closest approach – where it exerts the greatest gravitational pull on ring particles – shifts gradually, or precesses around the planet.

The up-and-down motion of Mimas relative to the ring plane causes vertical particle-density fluctuations known as spiral bending waves. These propagate inwards as a train of vertical corrugations up to 1 km in height. For a few days about the time of the Sun's passage through the ring plane (equinox), light strikes the ring particles at very low angles, and produces an exaggerated-scale three-dimensional view of the rings. The situation is similar to that when lunar craters and mountains are observed near the terminator – the day–night line – and everything is thrown into exaggerated relief. In the case of Saturn, the rings themselves are far too thin to be seen directly, and it is the vertical scale produced by the bending waves that allows ground-based observers to see the rings during the ring-plane crossings.

Voyager also confirmed that, far from being an empty void, there is a surprising amount of intricate detail within the Cassini Division. Already resolved into four bright bands in the far-encounter images, close-up views showed at least twenty ringlets and gaps. The ringlets vary from sharp-edged and opaque to diffuse and translucent, and of eight gaps definite enough to receive names, the most striking is the Huygens Gap, 417 km across, separating the Cassini Division proper from the sharp edge of the B ring at the 2:1 Mimas resonance position.

Obviously the tremendous amount of detail in the rings is far too much for resonance theory to explain. There are simply not enough

Detail of the intricate structure within the Cassini Division.

The tortured F ring, kinked by the small satellite Prometheus; its co-orbiting satellite Pandora appears as a dot at the lower right.

Spiral density wave structure in the A ring, produced by the 7:6 resonance with Janus.

resonances! In the B ring, where hundreds of bands appear, much of the structure remains unexplained. On the other hand, the resonance theory works fairly well in explaining structures in the A and C rings.

Thus the outer edge of the A ring corresponds to a 7:6 resonance with the co-orbital moons Janus and Epimetheus, and its shape, to a first-order approximation, is the seven-lobe figure predicted by theory; however, the density waves they produce are modulated by the once-every-four-year change of those satellites' orbits. The Pandora 5:4 resonance and the Prometheus 6:5 resonance produce significant features in the inner A ring. The outer A ring is dominated by spiral density waves produced by the ring moons Prometheus, Pandora and Atlas, which because they lie so close to the A ring have to complete many orbits before they realign with a given particle in the outer part of that ring. This gives rise to some rather exotic structures: for instance, the 36:35 resonance with Prometheus gives rise to a tightly wound series of spirals with 36 arms, with each several kilometres apart.

The moon Pan, only 28 km across, maintains the Encke Gap in Saturn's A ring by gravitationally nudging the ring particles back into the rings whenever they stray. Similar processes are thought to be at work as forming planets clear gaps in the circumstellar discs from which they form.

The Encke Gap is an outstanding example of shepherding by a moon – in this case, one embedded within the rings. The moon's existence was first suspected as early as 1985, when Jeffrey Cuzzi and Jeffrey Scargle, both of NASA Ames, noticed wavy edges in *Voyager* images of the Encke Gap. A year later, from a careful analysis of this structure, Mark Showalter and his colleagues were able to work out the orbit of the suspected moon, and in 1991 it was detected on *Voyager* archival images. It has been given the name Pan. The Keeler Gap, located very near the A ring's outer edge, has a complex inner edge, probably in part owing to a 32-lobe structure due to the Prometheus 32:31 resonance, but a more circular outer edge characterized by feathery wisps produced by another embedded moon, Daphnis, which was first visualized in *Cassini* images in 2005.

In *Voyager* images, the structure of the C ring resembles that of the Cassini Division, and consists of a series of narrow bright bands and narrow gaps. The gaps are produced by resonances, including a 3:1 resonance with Mimas. A striking feature is a very narrow, opaque, sharp-edged ringlet known as the Colombo or Titan ringlet; it is, in fact, a bending wave involving particles whose orbits are precessing like a top at the same rate as the orbital period of Titan.

The inner D ring is extremely tenuous, and presumably consists of material being lost to the rings through the process of angular momentum transfer. *Voyager* also confirmed the existence of a diffuse G ring between Mimas and the co-orbiting satellites Janus and Epimetheus, and the E ring, another diffuse ring whose densest part lies just inside the orbit of Enceladus, which consists of smoke-like particles on the order of only 1 micron across – so small that they are unable to scatter red light at all. Consequently, the ring has a bluish colour.

Observing the rings from behind the planet, the *Voyagers* found sharp instances of brightening of Ring C and the Cassini Division; the spokes; and the E, F and G rings in forward-scattered light. The brightening profiles showed that, at least for smaller-sized particles, the size varies across the rings. The particles had, of course, long been identified as dirty water ice. However, the *Voyagers* failed to clarify what dirtied the ice; they also failed conclusively to solve the problems of the rings' origin and ages. (Pristine rings would obviously be young rings, and dirty rings older.) After the *Voyagers*

The tiny moon Daphnis, only 8 km along its longest dimension, discovered by the *Cassini* Imaging team in May 2005, orbits in the Keeler Gap at the outer edge of the A ring, producing wavelike gravitational ripples as it passes. Because Daphnis' orbital inclination though small is not zero, these ripples have vertical relief, and cast shadows when Saturn is near its equinox.

had headed on into deep space, Jeffrey Cuzzi reflected on the implications of the fact that resonances with satellite cause the satellite to gain angular momentum and move outwards, while the ring particles lose angular momentum and collapse inwards:

Jeffrey N. Cuzzi, ring master at NASA's Ames Research Center.

> Theory predicts that an enormous amount of momentum must be flowing from spiral density waves in the A ring to their corresponding ring-moons. In fact, the transfer-rate should be large enough to collapse the A ring completely in fewer than 100 million years and, simultaneously, to hurl the ring-moons thousands of kilometres outward. The waves we see are actually much stronger than the existing theory can accommodate, yet the satellites now occupy orbits quite near the rings. It's certainly possible that the Saturnian system achieved its current state in the geologically recent past. But planetologists are understandably suspicious of such a conclusion.[21]

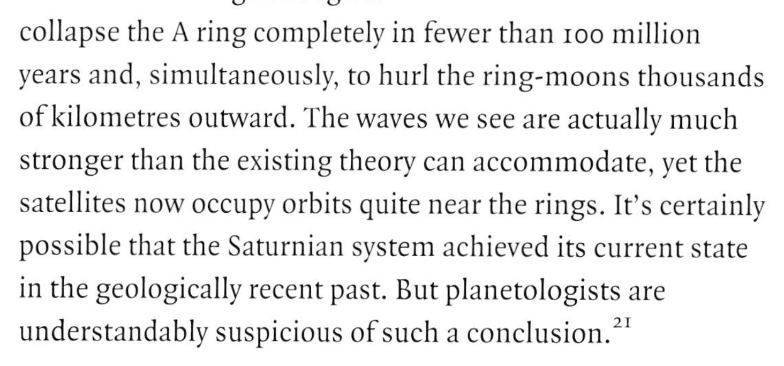

These were some of the burning questions *Cassini* was designed to settle.

FIVE

Cassini's Epic Mission

Although the *Voyager* missions of 1980 and 1981 were remarkable, both spacecraft were merely brief visitors of Saturn, passing by as tourists on their odyssey of the giant planets. The next stage in the exploration – Saturn, its rings and its moons – was to put a spacecraft in orbit round the planet. Known as *Cassini*, and organized as a complex collaboration between NASA, the European Space Agency (ESA) and the Italian Space Agency (ASI), it would be an ambitious (and expensive) flagship mission like the *Voyagers* themselves, *Galileo* to Jupiter and the *Vikings* to Mars. There were two main components: NASA's *Cassini* probe, a fully equipped observatory, which would enter into orbit round Saturn, and the ESA's *Huygens* lander, which would attempt a landing on Titan. The total cost of the mission was $3.26 billion (in 2000 dollars), of which the U.S. contributed 80 per cent, ESA 15 per cent and ASI 5 per cent.

Among *Cassini–Huygens*'s principal objectives were to study the following:

1. The three-dimensional structure and dynamic behaviour of Saturn's rings;
2. The composition of the satellite surfaces and the geological history of each body; in particular, the nature and origin of the dark material on the leading hemisphere of Iapetus;

One of the most iconic images from the *Voyager* missions: four days after its encounter with Saturn, *Voyager 1* looks back from a distance of 5 million km and images Saturn as can never be seen from the Earth, as an 'inferior' planet, that is, one lying closer to the Sun.

3 The three-dimensional structure and dynamic behaviour of Saturn's magnetosphere;
4 The behaviour of Saturn's atmosphere at cloud level;
5 The time variability of Titan's clouds and hazes, and the nature of Titan's surface on a regional scale; and
6 The composition, pressure and temperature of Titan's atmosphere, and the nature of its surface, as revealed from in situ observations.

Cassini set off in a spectacular night launch from Cape Canaveral aboard a Titan IVB/Centaur on 15 October 1997. (It did so without incident, despite the objections of a small number of protesters worried about the presence of the on-board nuclear reactor powering the spacecraft.) It made its way to Saturn via gravity assist fly-bys of Venus (April 1998 and July 1999), Earth (August 1999) and Jupiter (December 2000), with an incidental fly-by of asteroid 2685 Masursky en route to Jupiter. With the *Huygens*

Assembly of *Cassini–Huygens*. The *Cassini* orbiter (above) is being joined with the *Huygens* lander (below).

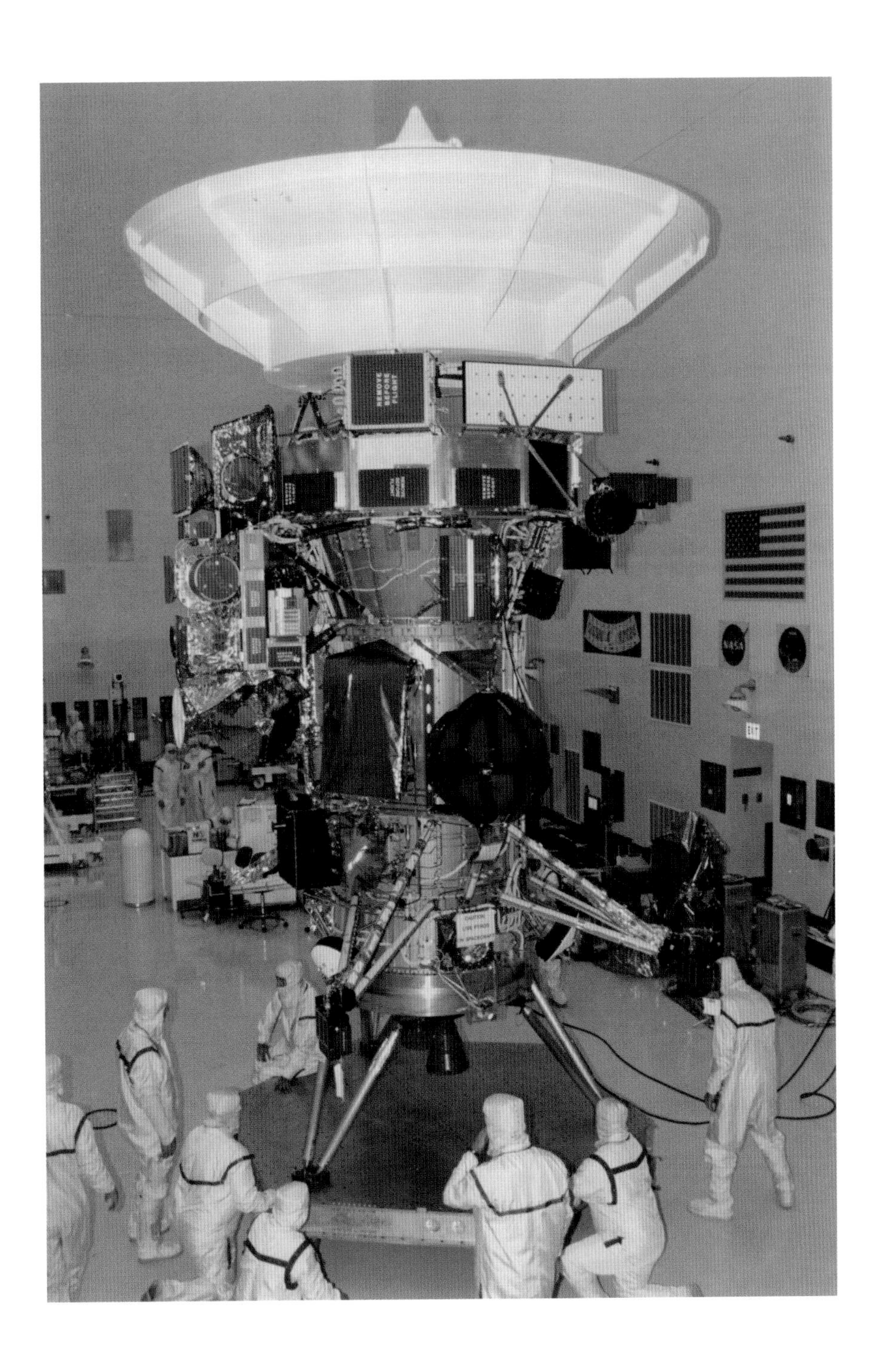
REMOVE BEFORE FLIGHT
NASA
EXIT

Cassini's launch on 15 October 1997.

lander still attached, *Cassini* successfully entered Saturn's orbit on 30 June 2004, during the height of the planet's southern hemisphere's summer, and began its surveillance. Meanwhile, the *Huygens* lander separated from its sister craft on 25 December 2004, and on 14 January 2005 dropped by parachute through Titan's atmosphere onto the surface, to achieve successfully the first-ever landing on a body in the outer solar system. Using the orbiter as relay, the lander sent information, including images, back to Earth for about ninety minutes.

Originally, *Cassini*'s mission was intended to last only four years, from June 2004 until July 2008. During this mission, called the Cassini Primary Mission, the spacecraft's orbits were varied in order to allow close-up views of the rings and inner moons. After completing the Cassini Primary Mission, and with the spacecraft showing continuing vigour and sending back vast amounts of

Cassini arrives in orbit at Saturn.

On *Cassini*'s arrival at Saturn, the northern hemisphere was in winter, as shown by this image captured by *Cassini* on 6 October 2004.

useful data about the Saturn system, a two-year extension (known as the Cassini Equinox Mission) was approved for funding. This allowed surveillance of Saturn about the time of the Sun's passage through the ring plane (August 2009). At the end of this extension, another extension, lasting seven years, was approved (the Cassini Solstice Mission). During this second extension, the spacecraft made 155 revolutions round the planet, 54 fly-bys of Titan and 11 fly-bys of Enceladus.

At last, after 13 years 2 months – or nearly a half Saturnian year – in orbit around Saturn, *Cassini* was finally running out of fuel. Having arrived during the height of the northern hemisphere winter, Saturn was now nearing the height of the northern hemisphere summer. *Cassini* was now directed to a close encounter with Titan and began the 'grand finale' stage of its mission, during which its elongated orbit reached below the inner edge of the D ring to only 3,000 km above the planet's cloud tops. After making a number of

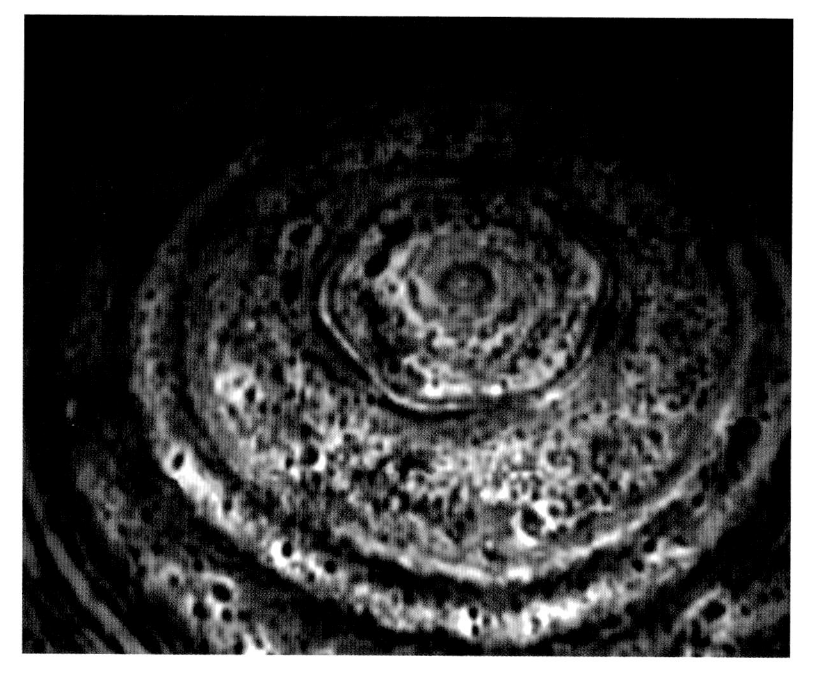

The north polar region, shrouded in night during the long northern hemisphere winter, as imaged in October 2006. The North Pole Hexagon appears as captured with *Cassini*'s infrared instrument, using Saturn's thermal glow at 5 microns as the light source. The Hexagon extends deep into the atmosphere, and is likely produced by a strong pole-encircling wave extending deep into the atmosphere.

Saturn, as it appeared in July 2008 soon after *Cassini* began its extended Equinox Mission, with the northern hemisphere, which was experiencing summer when the spacecraft first arrived, now moving into spring. The image was taken with the spacecraft located at a southerly elevation of 6°.

As *Cassini* passed into the shadow behind the planet on 17 October 2012, it captured this backlit view of Saturn and its rings.

daring passes through D-ring gaps, *Cassini* made a final pass by Titan, which sent it careening into Saturn on 15 September 2017 – its destruction intended to prevent accidental contamination of a potentially habitable moon with terrestrial microorganisms. Its fiery end in Saturn's upper atmosphere provided a dramatic finish to an epochal mission of discovery.

Return of the Spokes

On arrival at Saturn, one of the first discoveries made by *Cassini* was a negative: there were no spokes. At the time, it was generally believed that the spokes were always present, though perhaps

Night falls over Saturn's rings. Once every fifteen years, at equinox when the Sun is directly over Saturn's equator, night falls over the entire visible ring system for about four days. From Earth the rings are then seen nearly edgewise, but in August 2009 Cassini was 20° above the ring plane when it captured the series of images used to produce this mosaic.

The subtle structure of Saturn's rings – a natural colour view. *Cassini* image taken on 22 August 2009.

only visible when the observer lay close to the ring plane. Though they had not been seen even by the Hubble Space Telescope since October 1998, *Cassini* was expected to do better. However, it also failed to find them, even when close to the ring plane, and not until September 2005 were they finally recovered. Their visibility or invisibility was not, then, merely an observational effect: the spokes prove to be a true seasonal phenomenon, and disappear for periods of time, depending on the solar elevation above the ring plane.

Apparently, the spokes only form when the elevation of the Sun is less than 17° above the ring plane. (Notably, O'Meara's observations took place when the Sun was between about 3° and 16.5° above the ring plane – during a theoretical time of maximum spoke visibility. Though there are a few earlier drawings that appear to show spokes, none was made at the right season.)

As shown from the brightness of the spokes in views from behind the planet, they clearly consist of fine dust particles, with a narrow size distribution centred at about 0.6 microns. A possible explanation for their formation is that these tiny particles acquire electrostatic charge when they move out of the planet's shadow into sunlight. The charged particles are levitated above the ring plane into Saturn's magnetic field, co-rotating with the planet. Other possible explanations include micrometeorites striking the ring and whipping up a cloud of charge plasma (which then gets levitated above the ring plane and into the magnetic field), or even powerful lightning strikes that surge up from Saturn's clouds, strike the rings and blast out jets of electrically charged dust that appears as spokes. The latter theory has received additional support from spacecraft data showing that the spokes occur with about the same frequency as the planet's magnetic storms.

Even Saturn Gets the Blues

As noted above, when *Cassini* arrived at Saturn, it was near winter solstice in the northern hemisphere, and the atmosphere above the clouds was as blue as that of Uranus or Neptune. The finding surprised some of the scientists on the *Cassini* team, though it was not unexpected by at least some old hands at Earth-based observations. As *Cassini* continued to observe the planet through the Sun's passage through the ring plane (solstice) in August 2009, and as the seasons changed, with winter now passing to the southern hemisphere, the northern hemisphere's colour faded and the southern hemisphere became blue.

Cassini's observations demonstrate clearly what had been suspected from Earth-based observations: the blue winter hemisphere is a regular seasonal effect, owing to a combination, presumably, of the Sun's greater obliquity near the winter solstice and direct blocking of sunlight by the rings reducing ultraviolet

Looking like a view over Uranus, with its spider's thread rings, rather than Saturn, this *Cassini* image shows the ring shadows against the blue northern (winter) hemisphere of Saturn, on 18 January 2005. The satellite is Mimas.

light and thus decreasing the amount of overlying haze, as discussed earlier.

Rings Galore

Achieving a better understanding of the workings of the intricate rings, as well as their origins and evolution, was, of course, one of *Cassini*'s chief objectives. In contrast to the dramatic but cursory investigations of the *Voyagers*, *Cassini* was able to study the rings with multiple instruments from a wide range of observing angles and over a long enough period to detect changes due to shifts in the orbits of the moons hugging the rings (ringmoons) or clumping of particles within the rings themselves. Though the *Voyagers* had partly vindicated the classical resonance theory by showing that moons like Mimas disturb the orbits of ring particles and create

Illuminated side of rings, showing intricate structure (from left): the outer part of Ring B, Cassini Division and Ring A.

The backlit side of the rings, imaged by *Cassini* on 9 May 2007. This view is taken from above the northern hemisphere of the planet, which is still bluish above the shadow of the rings.

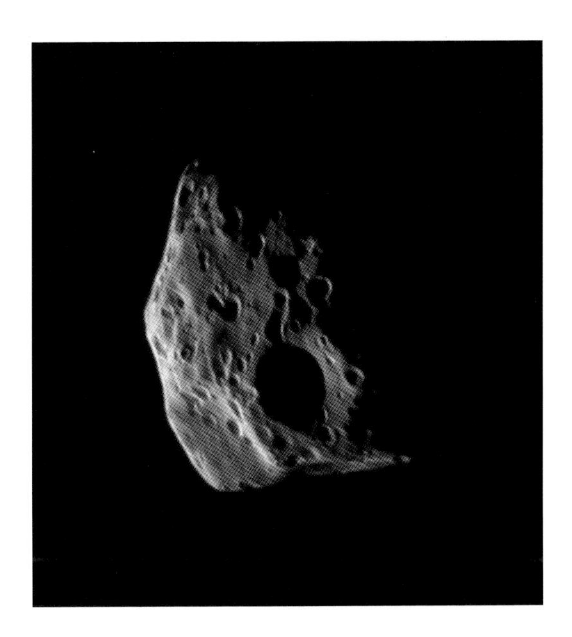

Epimetheus, imaged by *Cassini* on 30 March 2005, from a distance of 75,000 km.

resonance features such as the outer edge of the B ring, they also showed a much more complicated ring structure than anyone had imagined.

Satellite resonances ripple density waves across the main ring system, and also produce sharp edges like the outer edge of the A ring (the latter being produced by resonances with two co-orbital moons, Janus and Epimetheus). As noted, the structure of the A ring is generally well understood: thus, as noted earlier, the Encke Gap is produced by the embedded moon Pan, discovered from *Voyager* data, while another embedded moon, revealed by *Cassini* and named Daphnis, clears the Keeler Gap near the A ring's outer edge. Other bodies on the order of 100 m in diameter form propeller-shaped wakes as they migrate slightly inwards or outwards in gravitational interactions with nearby ring particles, attempting, but not quite managing, to clear out a gap. (Such bodies are referred to as 'moonlets'. An object not large enough to clear a continuous gap is a moonlet, while one that successfully clears a gap, like Pan and Daphnis, is considered a full-fledged moon.)

The Cassini Division shows an unexpectedly complex structure, including sharp-edged ringlets and gaps which are well enough defined to merit individual names (thus, for instance, the Huygens Gap marks the inner edge bordering the B ring). In the B ring, though much of the structure remains poorly understood even after *Cassini*, there are not only the usual gaps and ringlets but vertical relief: in one of the most stunning revelations of *Cassini*, made around the time of the August 2009 equinox, when sunlight fell at very low angles on the rings, the B ring's

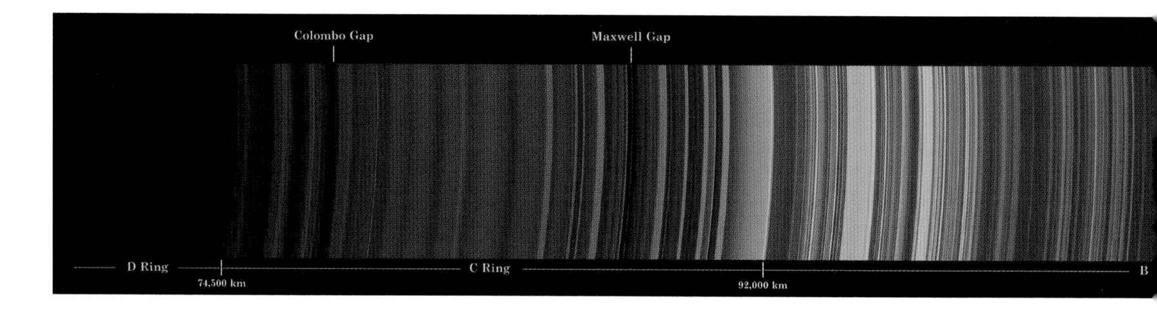

Dark side mosaic of Saturn's rings.

Prometheus by Saturnshine, *Cassini* image taken on 27 January 2010.

sharp outer edge showed a 20,000-km ridge of waves, with crests rising some 2.5 km above the ring plane and casting long shadows like those cast by the mountains of the Moon when seen along the terminator.

In the C and D rings, changes have been observed in the interval between the *Voyagers* and *Cassini*. An undulating spiral structure, some 19,000 km across and with peaks some 3 m high, extends across the entire D ring and reaches as far outward as the

inner C ring. It did not exist at the time of the *Voyager* fly-bys, but apparently, as inferred by Matthew Hedman and his colleagues, is an effect of a comet impact occurring as recently as 1983.[1] Such an impact would have tilted ring particles and caused them to precess, the inner ones faster than the outer ones. Because of the differential orbital periods of the particles, the disturbance would become ever more tightly wound over time. A similar pattern of more closely spaced ripples apparently records another comet impact, or more likely two impacts separated by about fifty years, occurring sometime in the fourteenth century, when Dante and Chaucer were still writing.

The F ring, despite its small mass and recent discovery, remains among the most intriguing of the rings. After *Voyager*, it had been believed that both of the ringmoons, Prometheus and Pandora, acted as ring shepherds. *Cassini*, however, showed that, in fact, they do more to stir up the motions of particles in this region than to stabilize it, with Prometheus, in particular, periodically plunging into it and drawing out streamers of fans and particles. Other bodies near the F ring observed by *Cassini* have such chaotic motions that they are hard to keep track of, and violently collide with the F ring's core. The real mystery of the F ring is not its twisted and braided character but the fact that the central core is as consistently shaped as it is, given all the factors that act to disturb it. The explanation for this was finally provided by Jeffrey Cuzzi and his colleagues, who showed that, apparently as a mere accident of the ring's position, Prometheus and particles in the F ring are in 'anti-resonance' – because of precession of Prometheus in its orbit, each gravitational tug by Prometheus cancels out its previous one, and this holds the F-ring particles in their orbit.[2] The effect is especially strong in the case of particles not perturbed by Pandora. Thus, contrary to previous expectation, Prometheus but not Pandora acts as a shepherd to the F ring.

In addition to its observations of the main rings, *Cassini* also made significant studies of the diffuse rings, composed primarily of particles less than 100 microns in size (and thus, while insusceptible to gravitational disturbances, they are sensitive to non-gravitational disturbances such as radiation pressure, magnetosphere interactions and the like). Even within the main rings, several narrow, dusty ringlets have been detected, while the aforementioned spokes consist of such small particles and are clearly involved in interactions with the magnetosphere. Beyond the F ring is the G ring, near whose inner edge is a bright arc of material that is likely confined by a 7:6 resonance with Mimas, and the diffuse E ring, whose densest part lies just inside the orbit of Enceladus and consists of smoke-like particles on the order of only 1 micron across – so small they are unable to scatter red light at all. Consequently, the ring has a bluish colour. It is of such vast extent that it encompasses all the icy satellites Mimas, Enceladus, Tethys, Dione, Rhea and – as first shown by *Cassini* – even Titan. Over time it ought to disperse were it not continually replenished by material from Enceladus' geysers (see Chapter Six). There is also the faint Phoebe ring, which is the source of the dark material that coats Iapetus' dark hemisphere.

For convenience, the following tables summarize information about the dimensions of the rings.

Diameters of the Rings				
Ring	Saturn radii	Kilometres	Miles	Earth diameters
D ring	2.472	148,983	92,573	11.679
C ring	3.054	184,059	114,369	14.429
B ring	3.902	235,166	146,125	18.435
A ring	4.538	273,496	169,943	21.440
F ring	4.652	280,367	174,212	21.979
G ring	5.80	349,554	217,203	27.403
E ring	16	964,288	599,181	75.593

Widths of the Rings				
Ring	Saturn radii	Kilometres	Miles	Earth diameters
D ring	0.126	7,594	4,719	0.595
C ring	0.288	17,357	10,785	1.361
B ring	0.424	25,554	15,878	2.003
A ring	0.242	14,585	9,063	1.143
F ring	–	varies 30–500	20–300	–
G ring	0.08	4,821	2,996	0.378
E ring	5	301,340	187,244	23.623

From Worlds to Rings: How Did the Rings Form?

Determination of the mass of the rings has always been one of the goals of Saturn studies and has important implications for the age of the rings, since it was only during the early history of the solar system (during the first 700 million years or so) that there was a sufficient quantity of debris careering about the solar system to have created relatively massive Saturnian rings. Estimates of the rings' mass based on *Voyager* data were rather uncertain, since during their fly-bys the two spacecraft always remained – quite intentionally – well outside the rings, where they felt the combined pull of both the rings and the planet. The best estimates were that the rings might have a mass on the order of 120 to 200 per cent that of Mimas.

In that case, they might be primordial – that is, they might have formed together with Saturn itself.[3]

It turns out, however, that the *Voyager* mass estimates were much too high. A definitive result was obtained by *Cassini*. Though during the first 12½ years of its mission the spacecraft orbited outside the rings and thus felt the combined pull of the rings and the planet, during the 'grand finale' phase it dived 22 times between the inner rings and Saturn's upper atmosphere. It thus felt the rings pulling in one direction and Saturn in the other, allowing the gravitational effects of the rings to be disentangled from those of the planet for the first time. The result was that the rings are only about 0.4 times the mass of Mimas, which means they are geologically quite young.

This is also evident from their brightness. Currently the ring particles appear to consist of 90 to 95 per cent water ice. At the time the rings formed, they were presumably almost pure ice, in which case they would have looked dazzlingly white, though they are now noticeably reddened, with both the A and B rings appearing, even to the eye, redder than any of Saturn's icy moons. Material in the C ring and the Cassini Division is even darker, with a colour more like neutral grey. The most straightforward explanation for these colours is that some other material has been mixed in with the ice, most likely interplanetary dust from comets and asteroids which would, over the 4.6-billion-year lifetime of the solar system, have added to the rings the equivalent of their original mass. This interplanetary dust consists of roughly one-third rock and one-third carbonaceous tars. When mixed with ice in equal proportions, this would cause the ice to appear dark and dirty and this is what we see, in fact, in the rings of the other planets – Jupiter, Uranus and Neptune, all of which are very dark. Accordingly, Saturn's rings should also appear dark if they were anywhere close to the age of the solar system.

Of course, the rate of darkening depends not only on the age but how fast dust falls onto the rings. This was measured by *Cassini*'s Cosmic Dust Analyzer, and proved to be some ten times greater than previously estimated. Combined with the low mass of the rings, this means not only that Saturn's rings are younger than the solar system, but much younger – perhaps 10 to 100 million years old. We are, then, seeing them while they are still quite young and fresh, their gloss not yet taken off by the relentless onslaught of incoming dust.

Their youthful age also significantly constrains theories of their formation. One plausible idea is that they formed by the impact of a comet into a small satellite once located near the Roche limit. Debris from a satellite outside the Roche limit could reassemble itself into a small moon, but any scattered inside would be unable to re-form and instead would continue to orbit the planet as ring particles. Such collisions were, of course, much more common during the early history of the solar system. However, they still occur from time to time. On Mimas, there is a crater, Herschel, that is large enough to have come close to shattering the moon altogether. Perhaps another, smaller moon was not so lucky – though, in that case, the rings ought to contain traces of the silicate-rich core of such a moon, which so far have not been detected. It is also possible that the moon Hyperion, with an irregular shape measuring 350 km along its longest dimension and looking very much like a collisional fragment from a larger body (its surface is oddly spongiform), took a blow from a comet, with pieces of its icy shell being tossed inwards to form the rings.

Other possible scenarios include the involvement of a rogue object from the Kuiper Belt, a region of icy objects which are mostly located beyond the orbit of Neptune. A number of Kuiper Belt Objects (KBOs) have been perturbed into eccentric orbits that bear them inwards towards the Sun, and a few even cross inside the orbit of Saturn. Collectively known as centaurs, these objects include 2060 Chiron, discovered in 1977, with a diameter of 272 km. At aphelion, Chiron approaches close to the orbit of Uranus; at perihelion, it crosses inside that of Saturn. Because of its close approaches to these planets, its future behaviour cannot be predicted, and there is a chance that a future close approach to Saturn will throw it on a hyperbolic orbit out of the solar system or – somewhat more probably – following an approach to Uranus, redirect it towards the inner solar system, where it might be captured by one of the other giant planets, Saturn or Jupiter, and become a member of these planets' large comet families. Phoebe, with its large, highly inclined, eccentric orbit, and retrograde motion (it travels round its orbit in the opposite sense to the main moons), is likely to be such a one-time Kuiper Belt Object captured by Saturn in the remote past. Though obviously objects as large as Chiron and Phoebe are rare, and there is almost no likelihood of any of these having made a recent strike on the Saturn system to form the rather lightweight and bright rings we see today, there are many smaller Kuiper Belt Objects which could have done so.

In fact, with the exception of the six largest moons, which probably formed from a circumplanetary disc round Saturn at the same time the planet itself formed, the rest of Saturn's more than 62 satellites are almost certainly bodies that have been captured or represent collisional fragments of the same. This is also true of Jupiter, with 79 satellites. The latter is an instructive case. In the spring of 2017, while searching for very distant solar system objects with the 4-m telescope at Cerro Tololo Inter-American Observatory in Chile as part of the hunt for a possible massive planet beyond Neptune, Scott S. Sheppard (Carnegie Institution), David Tholen (University of Hawaii) and Chad Trujillo (Northern Arizona University) discovered no fewer than twelve new moons of Jupiter when the planet passed through their search field. (It took over a year to obtain enough observations to confirm that the objects were moving round Jupiter, so the discoveries were only announced in July 2018.) Nine of the new moons are part of a distant outer swarm of moons that orbit in the retrograde, or opposite direction of Jupiter's rotation, and form three distinct orbital groupings, suggesting that

Phoebe, as imaged by *Cassini*.

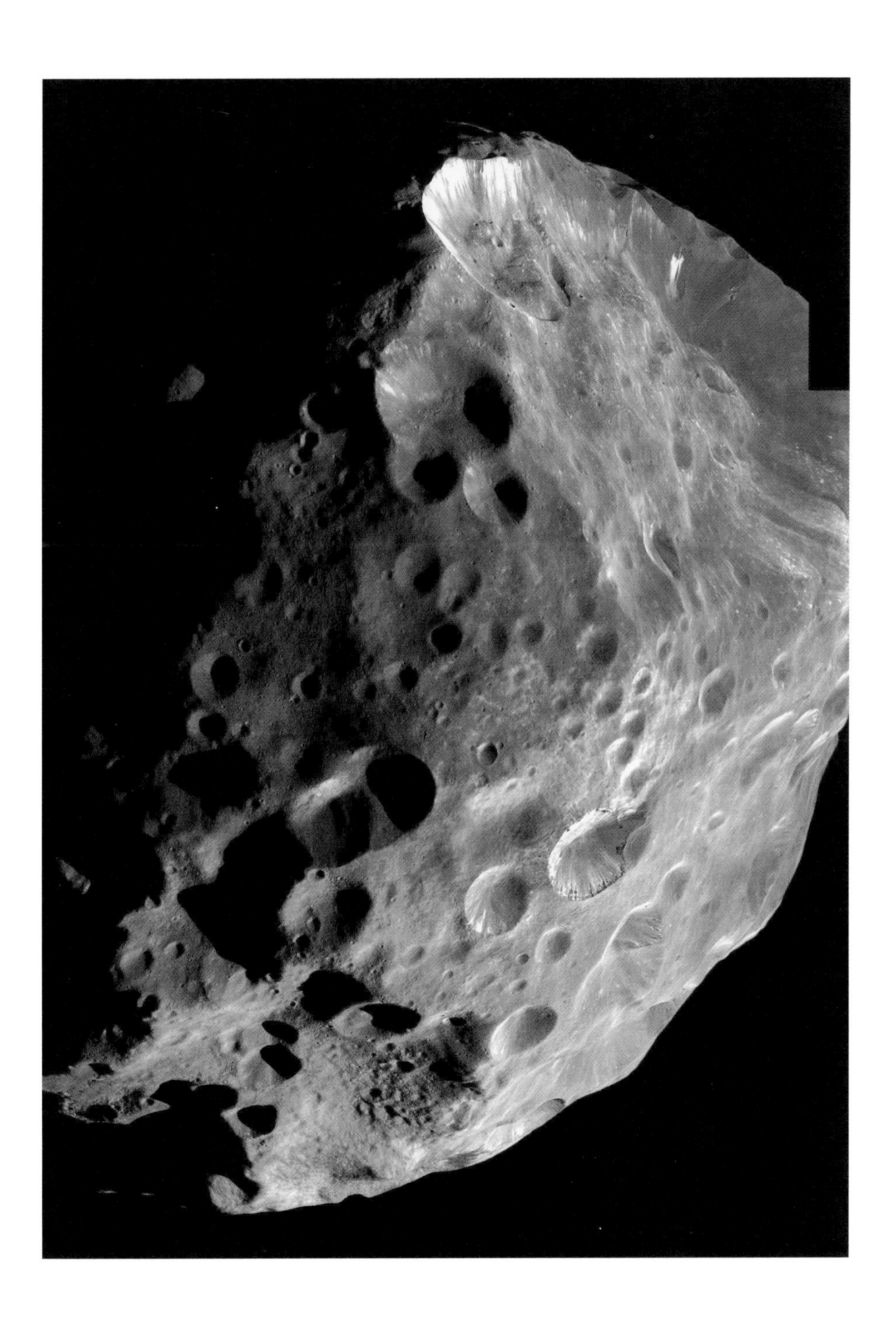

Cassini looks beyond Saturn, eclipsing the Sun, and catches a glimpse of a tiny distant world. It is sobering to think that of all the people who have lived on the Earth, not one has ever travelled farther than the distance between that pale blue dot and the Moon, which is just barely visible beside it.

three parent bodies were broken apart in collisions with asteroids, comets or other moons. Two are part of a closer, inner group of moons that orbit in the prograde, or same direction as Jupiter's rotation. They also appear to be fragments of a larger moon. The twelfth, however, is a decided oddity, in that though it, too, moves prograde, its orbit is highly inclined and crosses the outer retrograde moons. Obviously, as in traffic flow, wrong-way traffic is likely to end in grief, and eventual head-on collisions with other moons are inevitable. These collisions will quickly break the objects apart and grind them down to dust.[4] The several ringmoons outside the A ring likely represent fragments of larger parent bodies, and so almost certainly do the rings themselves.

The rings are not eternal. They are quite recent in cosmic terms, and what we see today reflects a highly evolved state from when they first formed. They will continue to evolve. Since they are continually losing material, they must have been much more massive in the past than at present, and based on current depletion rates as measured by *Cassini* during its passage through the gap between the rings and the planet before it plunged to its end in September 2017, they may be gone in only another 100 million years.

'Things fade,' said Alfred North Whitehead, and for all their sublimity, so shall the rings. Physically, they are a mere rubble disc, 'the natural end state of the collapse of a rotating cloud of debris'.[5] But as examples of the myriad rotating discs that are ubiquitous throughout the cosmos, from the spiral galaxies in which stars form down to the circumstellar and circumplanetary discs which give rise to planets and large satellites, they are also, in some sense, almost like Platonic ideas. They express those aspects Spinoza had in mind when he spoke of things seen *sub specie aeternitatis*, 'under the aspect of eternity', referring to what is universally and eternally true. As such, contemplating such things allows us mortals to escape, to a limited degree, the confines of our own brief spans.

As pure forms, if not as physical objects, the rings exist without relation to time. They are like the Platonic solids Kepler thought so much of. Standing on the steps of the observing ladder, looking into the eyepiece of a telescope, in the still cocoon-like expanse of the dome, with Saturn, its rings and its moons floating before the view, it is almost impossible not to feel something of what the American astronomer William Wilson Morgan (1906–1994) confided to the pages of his personal notebook:

> Ah – these hours of stillness – with supple brain – deep in the vistas of space, time, and Form – that Heavenly World of Form.[6]

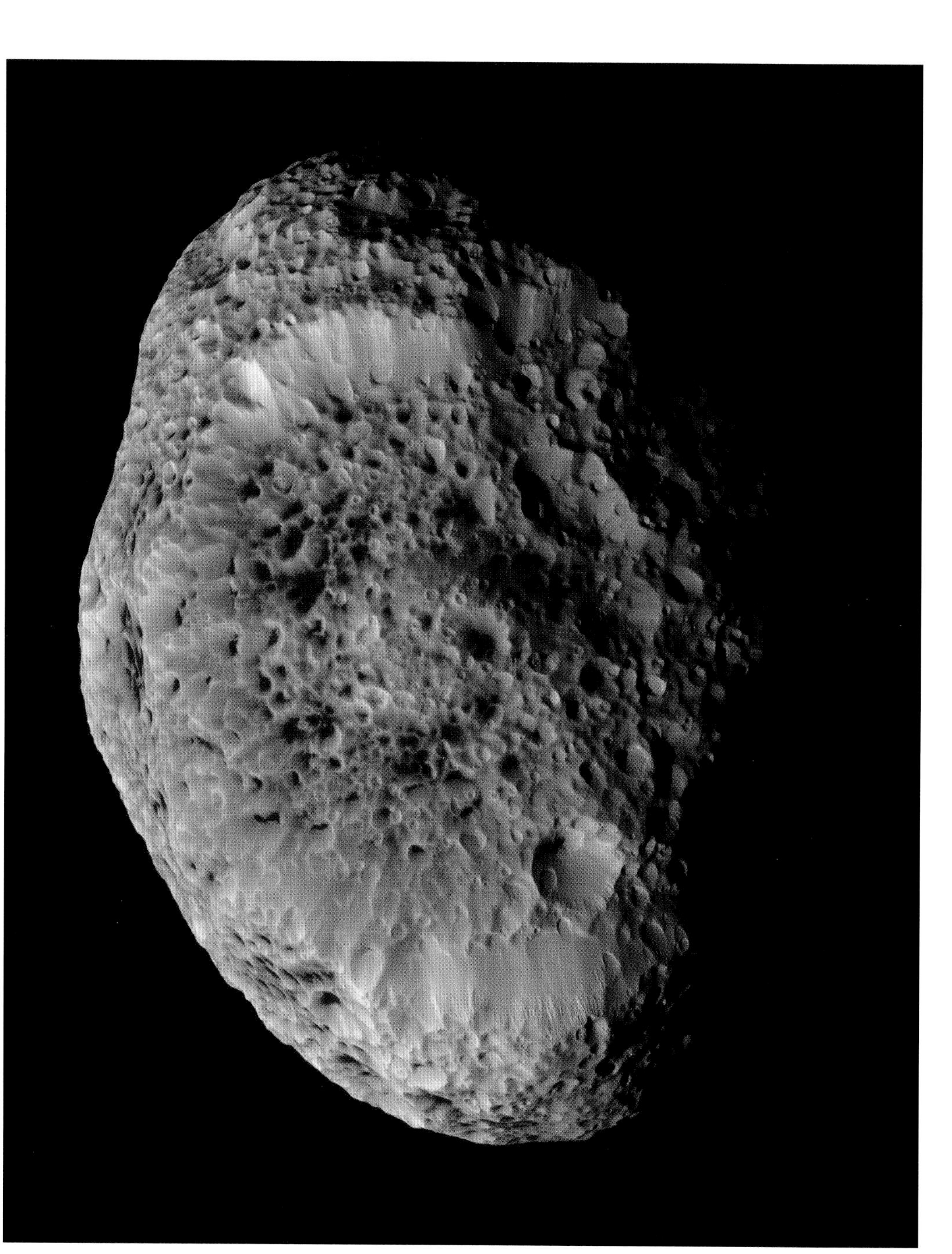

SIX

SATURN BY MOONLIGHT

Saturn's rings, as a swarm of particles – from aerosol-sized to several metres across – blend almost imperceptibly into a retinue of moons, of which, at latest count, 62 are known (and a 63rd suspected). So far, 53 have had their orbits worked out well enough to receive names.

Seven moons – Mimas, Enceladus, Tethys, Dione, Rhea, Titan and Iapetus – are thought to be primordial, that is, they formed at the same time as Saturn itself, presumably from satelletesimals that accreted from a circumplanetary disc in the same way the planets themselves took shape from the circumstellar disc round the Sun. Others, such as Hyperion, irregularly shaped and chaotically tumbling, and Phoebe, moving retrograde in a highly eccentric and inclined orbit, are no doubt captured bodies, possibly from the Kuiper Belt. The several dozen very tiny moons are, as described in the previous chapter, likely to be either small captured objects or fragments left over from long-ago collisions. This includes the ringmoons that hover just outside the outer edge of Ring A. Then there are the embedded moons that orbit inside the outer edge of Ring A and betray themselves by the gaps they clear within the ring, of which the most notable is 28-km wide Pan, which clears the Encke Gap.

The six largest satellites are truly worlds in their own right (see Appendix III), with interesting geological formations and, in the

Hyperion from *Cassini*.

case of Titan, an atmosphere. They exhibit phenomena, including transits, shadow transits and eclipses, similar to those exhibited by the Galilean satellites of Jupiter, though their phenomena are much rarer. The reason for this is the 26°.7 tilt of Saturn to the plane of its orbit, compared to Jupiter, where the tilt is only 3.1°. Thus, though Saturn's satellites, apart from Iapetus, revolve near the planet's equatorial plane, the tilt means that often they pass above or below the globe of the planet. Only for about two years, about the time of the ring-plane crossings, do windows occur during which satellite eclipses and shadow transits take place, of which those involving the largest satellites, Titan, Rhea, Dione and Tethys, are the ones of interest to amateur observers.

The first known observation of a Saturn satellite shadow transit was made by William Herschel on 2 November 1789, an observation not to be repeated until 15 April 1862, when William Rutter Dawes did so with a 21-cm refractor. That it took so long was not so much a matter of the difficulty of the observation or of Dawes's legendary eyesight – the observation can be accomplished with just a 10-cm telescope. Instead, it was due to a lack of predictions to alert observers, in combination with poor weather conditions at the critical times (as Dawes himself had experienced in 1848–9). Iapetus, with an 8° orbital tilt to the equatorial plane of Saturn, naturally marches to a different drummer; its phenomena are centred about two years prior to solar ring-plane passages, and include the rare instances when Iapetus is thrown into the shadow of the rings, of which the first observed, by Edward Emerson Barnard on 1–2 November 1889, led him to profound discoveries about the nature of the C ring. He wrote that

> the Crape Ring [sic] is truly transparent – the sunlight sifting through it; that the particles composing the Crape Ring cut off an appreciable quantity of sunlight; that these particles cluster more and more thickly – or,

The *Cassini* orbiter captured this spectacular image on 28 February 2005, from a distance of 2.6 million km from Saturn. The orbiter and satellites Dione (left) and Enceladus (right) all orbit almost exactly in the ring plane. This image shows, by the way, just how thin the rings really are.

in other words, the Crape Ring is denser as it approaches the bright rings.[1]

The Satellites from Spacecraft

In addition to studying the small satellites whose origins are obviously tied in with those of the rings themselves, the two *Voyager* spacecraft and *Cassini* obtained the first detailed images of the large satellites, and a great deal is now known about them.

As noted above, the six largest satellites are thought to have formed from circumplanetary discs, or subnebulae, similar to the circumstellar discs from which the planets themselves formed. These subnebulae must have formed early in the planet formation process, before the protoplanet had acquired sufficient mass to open a substantial gas gap in the surrounding nebula, and so managed to pull in gas to form a rotationally supported compact disc. The details are far from certain, but presumably satelletesimals formed from the gas in these discs and subsequently accreted into satellites.

The average densities of the satellites can be used to estimate the approximate proportions of condensed volatiles (mainly water ice) and rock (silicates and metals). The Saturnian system consists of one massive satellite, Titan, with a density of about 1,500 kg/m^3, very similar to the densities of Jupiter's Ganymede and Callisto. This density suggests a composition of about 50 per cent water ice and 50 per cent rock. The six mid-sized satellites – Mimas, Enceladus, Tethys, Dione, Rhea and Iapetus – on the other hand, have a wide range of densities, from 1,600 kg/m^3 for Enceladus to 991 kg/m^3

Mimas, with its large crater Herschel.

for Tethys, which doubtless represent actual differences in their compositions. They are, therefore, unlike Jupiter's satellites, which show systematic variations based on distance: thus the inner two, Io and Europa, are rocky objects, while the outer two, Callisto and Ganymede, consist of mixtures of ice and rock. At the moment, the compositional variations of the Saturnian satellites, as well as the apparent ice-rich nature of the rings and the inner co-orbitals and ringmoons, are not well understood.

In order of distance, Mimas lies closest to Saturn, just outside the A ring and within the E ring. By virtue of its insider position, it plays the leading role in sculpting resonance features in the rings, as Daniel Kirkwood showed long ago. The surface is icy and heavily cratered, and includes an imposing 130-km-wide multi-ringed basin, Herschel, that makes Mimas look rather like the Death Star from the *Star Wars* movies. The impact that formed it must have come perilously close to shattering the moon altogether, and might well have given us yet another set of rings.

Enceladus was already recognized from the *Voyager* images as perhaps the most intriguing of all the Saturnian satellites, with a surface that is smooth and youthful in appearance, suggesting major reworking in recent geologic times. Completing each orbit

Hubble Space Telescope image showing Titan and its shadow in transit across Saturn, 24 February 2009.

of Saturn in 1 day 8 hours 53 minutes, Enceladus occupies a 1:2 resonance position relative to Dione, whose orbital period is 2 days 17 hours 41 minutes. (As mentioned in Chapter Three, the resonance is not stable, as is the case with Io, Europa and Ganymede in the Jupiter system, but inherently unstable, and must eventually evolve towards a different set of orbital relations.) At present, the resonance causes Enceladus to experience strong tidal forces from the more massive Dione. As in the case of Jupiter's satellite Io, in resonance lock with Europa, tidal flexing and bending keeps the interior of Enceladus warm and allows ongoing geological activity to take place.

The surface is dominated by fresh ice, which makes Enceladus the most reflective body in the solar system. The albedo, indeed, is

View of trailing hemisphere of Enceladus.

A mosaic of *Cassini* images showing the north polar region of Enceladus. The two prominent craters above the terminator are Ali Baba and Aladdin. The Samarkand Sulci run vertically to their left.

the same as that of freshly fallen snow, 0.81. (Curiously, it is more reflective than the speculum-metal mirror in the telescope used by William Herschel to discover it, which was no more than 0.75.) With such a mirror-like surface, Enceladus absorbs very little warmth from the Sun, and even at noon its temperature never rises above 75 kelvin (−198°C). Beneath this icy rind, however, the tidal energy generates enough internal heat to create a global saltwater ocean, in places perhaps as deep as 50 km.

Because of the internal heat, large areas of Enceladus have been resurfaced in the quite recent past. There are regions of older cratered terrain, fractured owing to deformation of the surface since the craters formed, and smooth, craterless areas, marked by numerous small ridges and scarps. The most interesting area is a distinctive, tectonically deformed and obviously very youthful region near the south pole. Near the centre are four fractures bounded by ridges, referred to as 'tiger stripes', surrounded by blue ice, which

Cassini's Visible and Infrared Mapping Spectrometer (VIMS) instrument showed to consist of crystalline water ice. From these four fractures, more than one hundred geysers spout, producing a plume of ice particles and vapours extending hundreds of kilometres across the surface. Though most of this material falls back again, some of it escapes into space to form the diffuse E ring.

Tethys.

In order to analyse the composition of the plume material, *Cassini* passed through it a dozen times, and found that though the material was mostly water, there were also traces of organic molecules – carbon dioxide and ammonia.

These findings were undoubtedly among the most significant of the *Cassini* mission – and indeed in the history of solar system exploration – and suggested that the subsurface oceans of Enceladus possess all the components necessary for life to originate. The lack of sunlight, by the way, is no stumbling block; indeed, on Earth, rich ecosystems, the base of whose food chains consists of chemosynthetic microorganisms able to derive energy by converting carbon dioxide into methane, exist in the neighbourhood of ocean vents or 'black smokers'. Perhaps such ecosystems exist on Enceladus. Similar conditions, including a subsurface water ocean, may exist on Jupiter's satellite Europa, but its oceans are at greater depth, and thus more inaccessible than those of Enceladus. In light of *Cassini*'s results, Carolyn Porco asked, 'Did this small icy world host a second genesis of life in our solar system? Could there be signs of life in its plume? Could microbes be snowing on its surface?'[2] Because of *Cassini*'s results, Enceladus has emerged, ahead of Europa and even Mars, 'as the most promising, most accessible place in the solar system to search for life', and as such, as a prime target for future missions to Saturn.

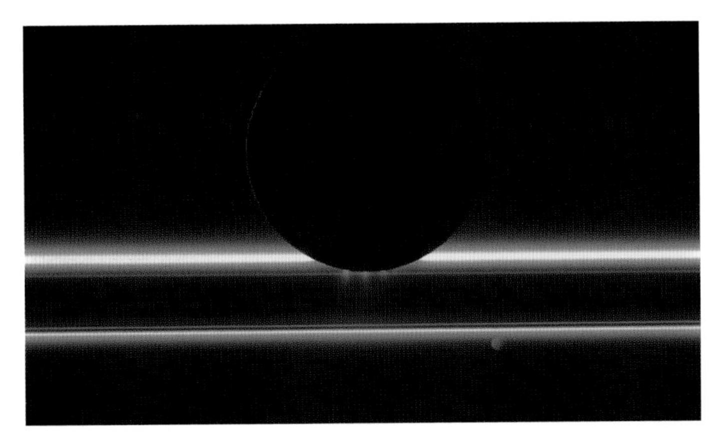

False-colour image of jets from the geysers in the southern hemisphere of Enceladus, taken from *Cassini* on 27 November 2005.

Continuing outwards, we next come to Tethys, 1,060 km in diameter, whose dominant feature, already known from *Voyager* images, is a large fracture, Ithaca Chasma, which stretches from one pole to the other and is the centre of a complex system of parallel

Rhea.

fractures. Also noteworthy is a 400-km-wide crater, Odysseus, whose floor contains a prominent peak.

The surface of Dione, at 1,120 km in diameter, and intermediate between Tethys and Rhea, was not visualized at close range by either of the *Voyagers* but showed only an intricate system of what appeared to be bright wisps, possibly surface frost, converging on a crater-like

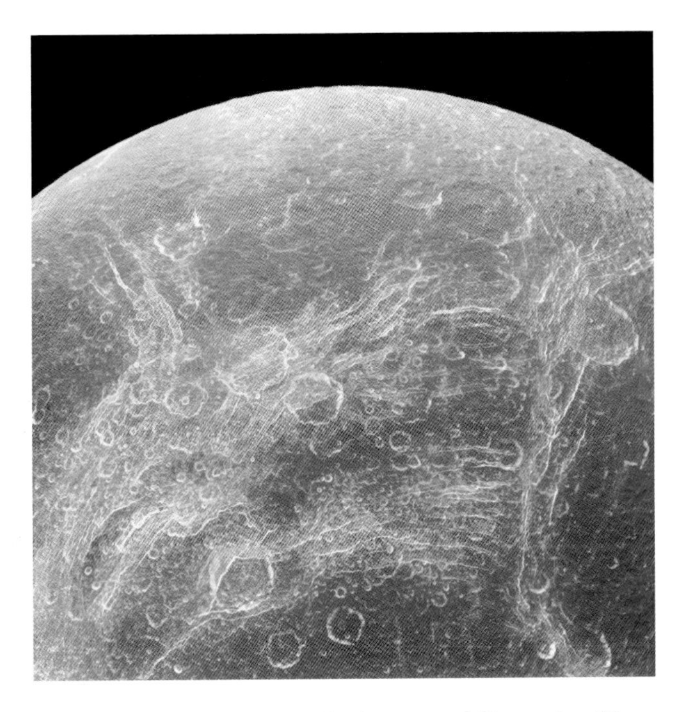

feature, Amata. The trailing hemisphere of Dione appeared smooth and dark, the leading hemisphere lighter and containing numerous impact craters, of which the two largest are Aeneas and Dido. Since Dione – like the other large satellites of Saturn – is tidally locked (that is, it always keeps one face towards Saturn, and the other away), it would be expected to have the highest cratering rates on the leading hemisphere, and fewer on the trailing hemisphere. (To make a rude analogy, the leading hemisphere would be rather like a windscreen moving head-first through a swarm of insects.) Since on Dione the reverse is found, it seems that it must once have been tidally locked to Saturn in the opposite orientation. A large impact may have changed the orientation; for that matter, the orientation might have changed multiple times. Meanwhile, the wispy features on the trailing hemisphere disclosed by *Voyager* prove not to be 'wisps' at all; rather, *Cassini* has shown them to be bright ice cliffs, hundreds of metres high, located along fractures (chiasmata).

Dione.

With Rhea, which has a diameter of 1,530 km, the leading and trailing hemispheres show the more typical and expected pattern. The leading hemisphere is more heavily saturated with impact craters than any other surface in the solar system, while the trailing hemisphere is dark and smooth. There are two impact basins in the 400- to 500-km range, of which the more northerly and less degraded is called Tirawa. There is also a prominent rayed crater called Inktomi, 48 km in diameter, which is one of the youngest impact craters in the Saturn system.

We will hold off on Titan for the moment, and first consider Iapetus, 907 km across and the only one of the large satellites that lies farther from Saturn than Titan. G. D. Cassini's three-hundred-year-old theory to explain the brightness difference of the two

Cassini mosaic of Iapetus, showing the bright trailing hemisphere. The large crater Engelier is near the bottom, and superimposed on and partly obliterating the older crater Gerin.

Iapetus, comparing the dark and bright hemispheres.

hemispheres was confirmed by the *Voyager*, whose images showed that the trailing hemisphere is ten to twenty times more reflective than the leading hemisphere. The latter is dominated by a broad black swathe extending across 220° of longitude and 110° of latitude. Observations from the Spitzer Infrared Space Telescope and *Cassini* have added further detail: the broad black swathe contains material from the outer (presumably captured) moon Phoebe. Solar radiation and micrometeoroid bombardment have been loosening dust-grain-sized material from its surface for billions of years, to form an extremely sparse 'Phoebe ring', larger and more tenuous than any of the other rings of Saturn. Each time Iapetus passes through this material, it collects dust particles on its leading hemisphere, which darken the surface; the other, trailing hemisphere is 'normal', covered with water ice. The tonal differences are stabilized by the fact that the dark hemisphere absorbs more heat and is warm enough so that any time water attempts to condense and freeze onto this surface, it sublimates away; water ice forms only on the brighter, cooler side. So each hemisphere retains its characteristic complexion.

Titan: A Planet-sized Satellite

We turn, finally, to Titan, by far the largest of Saturn's satellites – ten times larger than Enceladus, for instance – and the only one constructed on the grand scale of the Galilean satellites of Jupiter. (As noted earlier, it is surpassed only by Ganymede.) The only satellite in the solar system known to have an appreciable atmosphere, it was, for obvious reasons, a principal objective of the *Cassini–Huygens* mission.

Voyager images had shown that Titan is shrouded in a thick orange-brown smog of hydrocarbons produced by photodissociation of methane. The smog interferes with the ability of sunlight to reach the surface but at the same time allows thermal infrared radiation from the surface to escape into space – thus producing what is referred to as an 'anti-greenhouse effect', in which warming by greenhouse gases, such as methane itself, in the lower atmosphere is blunted. The atmosphere of Titan, therefore, is only weakly efficient as a heat trap; the temperature at the surface, at 94 kelvin (–179°C), is a mere 7° warmer than if Titan had no atmosphere at all.

Because of the smog, the *Voyager*s failed to make out anything of the surface, though Earth-based radar and Hubble Space Telescope observations in the infrared revealed patches with different reflectivities, which were identified as disconnected seas or lakes of liquid methane or ethane. Not until *Cassini*, however, did Titan more fully yield its long-held secrets.

Cassini imaged Titan's surface at near-infrared wavelengths, where the smog was most transparent, and also used radar to map the surface as had previously been done by the *Magellan* spacecraft in orbit around cloud-shrouded Venus.

Cassini radar mapping of Titan showed complex valley systems formed by the flow of liquid methane and ethane, and permanent hydrocarbon lakes near the poles. Because Titan's surface, at 94 kelvin, is so bitterly cold, both methane and ethane remain liquid and pool on the surface. The boiling point of methane is 111 kelvin

Smog layers over Titan.

(–161°C), and of ethane, 184 kelvin (–89°C). While Earth's surface is dominated by the water cycle, on Titan water is as hard as rock and instead has a methane cycle – hence its very striking, if illusory, Earthlike appearance, which includes rivers and river deltas. One of the latter – at least to me – looks rather like Long Island and the Hudson River Valley.

Many of the lakes on Titan are permanent, and, for reasons not yet entirely clear, tend to be found predominantly around the north pole. Some are impressively large; for instance, a methane lake at Titan's south pole is the size of Lake Ontario (and so has received the name Ontario Lacus). There are some with marshy edges, while others appear bone dry. They undergo seasonal changes, in which

Specular reflections from ethane lakes on Titan.

Titan's seasons resemble those of Saturn itself. Other surface features include vast dunes stretching about the equatorial region of Titan, where the 'sand' is composed of dark hydrocarbon grains whose appearance has been compared to that of coffee grounds. Despite the exotic material of which they are formed, their tall, linear shapes resemble those of terrestrial sand dunes, such as those found in the Namibian desert in Africa.

Radar image showing surface of Titan.

Cassini's numerous gravity measurements revealed that Titan likely harbours an ocean of liquid water (likely mixed with salts and ammonia) below the 'rock' forming the surface. This lies at a depth of 55 to 80 km. Thus, along with a handful of other worlds in the solar system, including Jupiter's Europa and Saturn's own Enceladus, Titan is a possible abode of life. The rivers, lakes and seas of liquid methane and ethane on the surface may even harbour life of some kind – though hardly life as we know it.

Meanwhile, detached from *Cassini* soon after the latter's entry into orbit, the probe *Huygens* began its mission to parachute through the atmosphere onto the surface of Titan itself. Using an on-board gas chromatograph/mass spectrometer, *Huygens*, on its descent, was able to analyse the composition of the atmosphere, and register changes in the intensity of its radio signals in order to measure the speed and direction of its winds. Further, the probe's camera imaged the surface at various heights as well as from the landing site on the surface itself.

The descent lasted 2 hours 27 minutes, and touchdown occurred in the equatorial, dune-covered plains. Hitting the ground with a speed similar to that experienced by a ball dropped from a height on Earth of about a metre, it bounced, slid, wobbled and kicked up a cloud of dust before finally coming to rest. The landing site lay near a series of branching drainage channels that appeared to drain from an adjacent light-coloured region into a darkish sea; subsequent analysis showed that the probe had, in fact, landed within this sea. Images taken from ground level indeed showed a flat plain covered

Area of branching drainage channels in *Huygens*'s landing area.

with rounded pebbles, which recalled scenes from other worlds such as those taken by the *Venera* probes on Venus or the *Viking*, *Pathfinder* and *Opportunity* and *Curiosity* landers on Mars – or even basaltic lava fields on Earth, like those on the island of Hawaii. The skies appeared the same light russet that one sees when low winter sunlight filters through a cold haze.

Appearances can be deceptive; the analogies to terrestrial scenes were illusory. Instead of basaltic lava fields, the pebbles (none of which was more than about 15 cm across) were water ice, hard as rock, and presumably coated in hydrocarbons. The russet sky colour was due to the much greater attenuation of blue light compared to red light by Titan's hydrocarbon smog. Even during the daytime, the skies on Titan are crepuscular but not dark; at the time of *Huygens*'s landing, the illumination level was approximately the same as that ten minutes after sunset. This was still enough to cast sharp, low-contrast shadows. The probe had little more time than just 'to look about it and to die', to adapt Alexander Pope; within 72 minutes of landing, its battery gave out, and the probe went silent.

One thinks at this point of Christiaan Huygens with his 50× refractor and who, more than three and a half centuries ago, first noticed Titan as a faint star. What progress since then! Now the probe named for him, a small instrument-laden package, our emissary, has unveiled it as a world – admittedly, one of the strangest, most unearthly that we know. *Huygens*, though long since out of contact, remains standing among the rocks of ice and the lakes of ethane and methane which make up the landforms of that strange alien world, fronting coastlines that are the most remote to which our curiosity and ingenuity have so far led us.

Close-up Views of the Ring Moons

In our survey of the rings, we have often encountered several tiny moons located outside, or in two cases, inside, the outer edge, which play an outsized role in sculpting the rings – for instance, Prometheus, which shepherds the particles in the F ring; Pan, which clears out the Encke Gap; and Daphnis, which moves along the feathery outer edge of the Keeler Gap.

An unexpected bonus of the 'grand finale' phase of the *Cassini* mission was that, in addition to making ring-grazing passages to study Saturn, its rings and magnetosphere, the spacecraft also made close flybys of five of the ring moons – Atlas, Daphnis, Epimetheus, Pan and Pandora – allowing them to be examined in detail. Their surfaces were found to be highly porous, suggesting that these moons have formed in stages, presumably by collision of original denser cores (perhaps remnants of the grand catastrophe that formed the rings) with ice and dust from the rings that has stuck to them. This accreted material has produced little skirts round their equators, giving them ravioli-like shapes. The surfaces of the two moons within the A ring's outer edge, Atlas and Daphnis, are the most altered by ring material, and appear reddish, the colour of the rings – though neither of these moons is orbiting in a cloud of ring material right now. The moons located outside t he A ring are more bluish, showing evidence that they have been coated with material originating in Enceladus' similarly coloured icy plumes.

According to *Cassini* team member Bonnie Buratti, the same process is doubtless going on throughout the rings, with the largest

Cassini looks back as the Sun disappears behind the ball of Saturn and beyond the diffuse G and E rings.

ring particles also accreting ring material around them: 'the rings and these moons are really the same kind of object – the rings are made of small particles, and these moons are the biggest versions of these particles'.[3] No doubt close examination would show them to have the same blob-like, skirted ravioli shapes of their larger exemplars. So Saturn's rings and moons, interacting as part of a dynamic system, give posthumous existence of a shattered moon.

SEVEN

Observing Saturn

No matter the telescope, Saturn is always worth a look. Its beauty always elicits gasps of awe from even casual viewers, and the varied phenomena of rings, globe and moons make it endlessly enthralling.

Its globe is remote and its features understated compared to Jupiter's, and too small to be an easy object. Even at its maximum apparent diameter, Saturn is never more than the size of the crater Copernicus on the Moon, and that includes the rings! It therefore takes a fair amount of practice and dedication to get beyond the initial stage of lovely but somewhat static revelation. Only years of persistent study can reveal such time-variable phenomena as the planet's changing ring aspects, seasons, and satellite and satellite-shadow transits and eclipses.

What you see depends on many factors – including the quality of your telescope and selection of eyepieces, the state of Saturn's atmosphere and the ring angle, the state of Earth's atmosphere (how turbulent it is) and your experience as an observer. What follow are some tips on how best to see what is arguably the solar system's most beautiful planet.

Experience. There is only one way to gain experience, and that is to look at the planet night after night; the more you look, the more you have the potential to see, especially as you become increasingly familiar with the planet's subtle details. Once orientated to the general appearance of the planet, you can start to home in on finer details.

Atmospheric 'seeing'. It is important to keep records that include the start and end times of the observation, date, location, size and type of telescope used and magnification employed. If any of this information is missing, the record's usefulness is compromised. Also important is an estimate of the stability of Earth's atmosphere (the 'seeing') on a scale ranging from 0–1 to 10. At 0–1, the conditions are the worst possible – the image of the planet is boiling, quivering or undulating badly, and it is probably just as well to put the telescope aside and wait for a better night. At 10, the planet appears 'rock steady', like an etching or steel engraving: 'perfect seeing'.

Remember that the Earth's atmosphere is in constant flux, so do not be discouraged if your first view of Saturn shows the globe and rings and little else. You will eventually learn how to appreciate and enjoy a night of perfect seeing, when the planet becomes a revelation. Specific locations have their own patterns. In Flagstaff, where the author observes, the 'monsoon season' – which runs

The northern face of the rings imaged on 6 June 2018, by the Hubble Space Telescope. The planet was approaching opposition, which it reached on 27 June, so only a very narrow sliver of the shadow of the ball on the rings is visible. Note the bright cloud in high southern latitudes. The structure of the rings visible in this image is more than can be seen from Earth.

from July until mid-September – is usually hopeless, because of the unstable condition of the atmosphere (with frequent clouds and thunderstorms). On the other hand, 'rock steady' seeing is not uncommon at other times of year. Horsetail (high cirrus) clouds often presage good seeing. In addition, sky transparency – the darkness of the sky – is not necessarily related to good planetary viewing. Often the atmosphere is settled by fog or smoke, and this is also often the case because of the 'heat bubble' over large cities. Obviously, observing on tarmac (or from below ground) is ill-advised, and one ought to take the telescope out, if it is stored indoors, or open the dome or run-off shed early enough to allow the telescope to come into thermal equilibrium with the air. This is especially the case with a reflector of the Newtonian type, where the tube is open.

Telescope and eyepieces. Saturn and its rings can be seen and appreciated through telescopes of all sizes using magnifications as low as 30×. (Remember, Huygens's telescope magnified only 50×, and Cassini made out the division named for him with one magnifying only 90×.) For serious studies, a 76-mm refractor or 154-mm reflector are probably at the low end, but will reveal the

Saturn, observed on 30 November 1874 by the French astronomer and artist Étienne Léopold Trouvelot, with the 38-cm Merz and Mahler refractor at Harvard College Observatory. Note distortions of the edge of the shadow of the ball on the ring.

reddish-brown hue of the major dark belts on the globe, and, depending on the tilt of the rings, a glimpse of the crêpe ring. Much more becomes visible in telescopes of 200 mm and larger – given the right conditions and investment of enough time. It should be pointed out that, though it is always recommended to use no higher magnification than the seeing permits – if the image becomes soft, it is best to go to a lower magnification – because of its low surface brightness, Saturn tolerates higher magnifications than, say, Jupiter; 20× per centimetre of aperture is a reasonable figure.

A List of Things to Look For

On scanning the planet under good seeing with a small- or moderate-sized telescope, attention should be given to the following:

1 The shape of shadows of rings on the globe and of the globe on the rings. Irregularities should be noted, such as distortions of the shadow edge, which can cause the shadow to appear abruptly cut off, with its curvature reversed at the Cassini Division or its outline displaced at the point where the shadow crosses. The effect is produced when the shadow edge strikes the Cassini Division at a narrow acute angle, causing a blurring similar to the black drop effect that occurs at the interior contacts of transits of Mercury and Venus. Another optical phenomenon worth looking out for involves the appearance of a bright area in the sunlit rings adjoining the edge of the globe's shadow – known as Terby's white spot, after the Belgian astronomer F.J.C. Terby (1846–1911), who first described it in 1889. The effect is purely an illusion, due to the eye's enhancement of contrasts between adjoining areas of different intensity, and because image

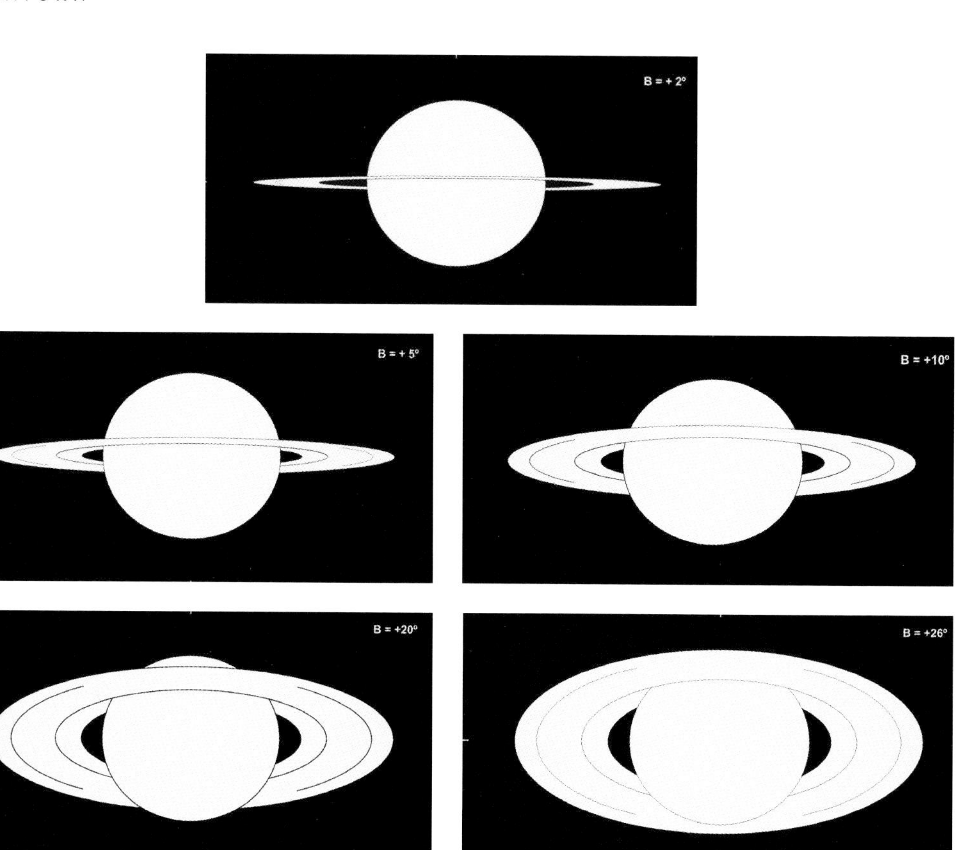

Saturn blanks for various ring aspects, to assist the observer in drawing the planet at different ring tilts.

processing routines mimic the eye–brain system, it can also appear on CCD images. Nevertheless, it can be striking.

2 The ring tilt is important. The more open the rings, the easier it is to appreciate the ring components and divisions. As the rings narrow, detecting even the Cassini Division can become a challenge. When the rings are edgewise, specialized observations – such as of the 'condensations' along the ring – can be made. Also the satellites are more easily detected, both because of the decreased glare of the rings but also because they then appear to shuttle back and forth along the ring plane, and so are easily distinguished from field stars.

3. The time from opposition is important. For making out fine divisions, it is best to avoid times too close to opposition, and instead to wait until there is a marked shadow of the ball on the rings. The maximum shadow occurs at quadrature, when the Sun–Earth–Saturn angle is 90°.
4. The number, colouring and positions of bands should be noted, and identified according to the standard BAA nomenclature. Any projections from the belts or bright or dark patches should be noted; rotation periods can be derived by timing transits across the Central Meridian. These measurements have been historically useful in plotting wind speeds on Saturn. In addition, observers should be on the lookout for whitish spots. Though after the outbreak of the Great White Spot of 2010–11, another on that scale is not expected for decades, smaller white spots are not uncommon; there was one, for instance, in very high latitudes in June 2018, which was captured by the Hubble Space Telescope.
5. Seasonal phenomena, such as the blue hemisphere that characterizes the autumnal and winter aspects, should be noted. In the rings, the spokes should be looked for, but they form only when the elevation of the Sun is less than 17° above the ring plane, and disappear when the rings are wide open. Sightings of the spokes at other times must be credited to illusion.

Drawing Saturn

Saturn is very challenging to draw, even for those with artistic talent, and quite impossible for those without. Though in the past the best solution was to produce stencils from card stock for representing the ellipses of the ball and rings, most observers nowadays will find prepared blanks, which are available from the British Astronomical Association.

In general, it is probably better to make partial sketches that focus on specific detail, such as spots or ring divisions, than to attempt to draw the whole planet. For the artistically inclined, and particularly for those who are experienced in the use of colour, the potential exists to create exquisitely beautiful renderings, such as those made by such skilful artist-astronomers as Trouvelot, Bolton and Abel. Though, unfortunately, drawing the planets is becoming a lost art, there are still some observers who are willing to take the time and effort, and doing so provides an indispensable link to the great era of visual amateur observers of Saturn and the other planets.

Imaging Saturn

CCD imaging has, of course, largely replaced visual observing of Saturn. The scope for useful work here is very wide indeed, but the technicalities are beyond the scope of the present book. Most experienced imagers use colour filters to enhance the visibility of the planet's atmospheric features. A light blue filter (Wratten 80A or 82A) will increase the apparent contrast between the belts and zones without significantly decreasing the brightness of the image. An orange or light red filter (Wratten 21 or 23A) darkens the bluish polar regions, while a light magenta filter (Wratten 30) may be helpful in increasing the visibility of many low-contrast markings. In addition, different filters can show clouds at different altitudes. Thus, as described in Chapter Three, there were multiple early CCD detections of the Great White Spot of 2001, which – even though the *Cassini* spacecraft was in orbit round Saturn at the time – provided invaluable data points during the period when no spacecraft observations could be obtained.

Advanced amateurs may carry out specialized observations. Methane band filters (such as the Baader Planetarium Methane

Filter 1.25") are sensitive to part of the spectrum in the infrared to which the human eye is not sensitive. In images taken with such filters (and infrared-sensitive cameras), the Great White Spot appeared dark, but images taken with an ultraviolet filter showed it bright. Analysis of such observations provided valuable information on the vertical structure of the Great White Spot.

Occultations of Stars by Saturn

Occultations of stars are most likely, of course, when Saturn is in a part of the zodiac rich in telescopic stars, such as the Gemini–Cancer region for northern hemisphere observers and Sagittarius for southern hemisphere observers. However, even they are quite rare – during all of 2018, even as Saturn was passing through the star-rich fields of Sagittarius, there was only one occultation by Saturn of a star brighter than ninth magnitude. Predictions of such events are provided by observations coordinated by the International Occultation Timing Association (IOTA). The observer should time as accurately as possible the immersion behind and emersion from each ring, and estimate any light or colour changes noted, as such events provide unique opportunities to probe the fine structure of the rings. Occultations involving bright stars – such as fifth-magnitude 28 Sagittarii on 3 July 1989, which was the brightest star occulted by Saturn in the twentieth century – are spectacular, but, of course, extremely rare. Indeed, the next occultation of a star brighter than sixth magnitude will not occur until 7 April 2032.

Satellites

A 50-mm telescope will show Titan, and a 75–100-mm telescope the next four satellites of Saturn. Titan's orange tint is evident, and so is the fact, first noticed by G. D. Cassini, that Iapetus appears two magnitudes brighter on the western side than on the eastern side of

Saturn. Enceladus, because it is so close to Saturn, needs at least a 15-cm and probably a 20- or 25-cm aperture; Mimas is even more difficult to see – though surprisingly Hyperion, despite its late date of discovery, is easier than either, but usually overlooked because it orbits so far from Saturn. It can be identified most easily when it is close to one of the other satellites – especially Titan. Ephemerides giving the positions of Saturn's satellites are published annually in the *Handbook of the British Astronomical Association*, the *Observer's Handbook of the Royal Astronomical Society of Canada*, the *Astronomical Almanac* and other publications.

Eclipses and Other Phenomena Involving Iapetus

Curiously, the first opportunity to probe the rings in detail involved an eclipse, not an occultation. This was E. E. Barnard's observation of the 1–2 November 1889 eclipse of Iapetus, which allowed him to establish the hitherto only-suspected transparency of Ring C. At intervals of fourteen or sixteen years, two years before the Earth's passage through the ring plane, the orbit of Iapetus becomes edge-wise to the Earth, and Iapetus appears to move in a straight line across Saturn, and transits, eclipses and occultations can occur. The next set of Iapetus occultations, transits and eclipses will occur in 2022–3 as follows:[1]

2022	
21 March	Occultation
29–30 April	Transit
8 June	Eclipse
9 June	Occultation
17 July	Shadow transit
17–18 July	Transit
27 August	Eclipse

27 August	Occultation
5 October	Shadow transit
15 November	Eclipse
23 December	Transit
24 December	Shadow transit
2023	
2–3 February	Occultation

This table is presented as something of a challenge to observers. As of the time of writing (August 2018), there do not appear to be any reports of transits of Iapetus or its shadow made by amateur observers from Earth, even though they should be almost as readily visible as those of Rhea.

The other satellites also undergo eclipses, occultations and transits (both of the satellites and their shadows), during the approximately five years out of every fifteen in which the apparent tilts of the rings and satellites are small enough for phenomena to be observable. Apart from those involving Titan – whose shadow is 0.8″ of arc across and very dark – they are rarely observed. Nevertheless, they are within reach of amateur instruments; the author has seen both Rhea's and Tethys' shadow transits with modest instruments.

There are also mutual events – satellite–satellite occultations and eclipses, which take place within one to two years of each ring crossing. The first such observation of this sort seems to have been an occultation of Enceladus by Tethys, observed by Robert Aitkin of the Lick Observatory on 11 November 1907. The eclipse of Rhea by Titan on 8 April 1921 was witnessed by several observers, but only in 1979–80 did sufficiently accurate and complete predictions become available. Since then, these phenomena have been routinely observed and imaged. The next mutual-event series will commence in 2024.

Edgewise Rings

The edgewise passages, in which the Sun and Earth cross the plane of the rings, occur at alternate intervals of 13 years 9 months and 15 years 9 months, during which the Earth passes once or three times through the ring plane. A list appears in the table titled 'Passages of the Earth and Sun through the Ring Plane of Saturn' in Chapter Two. Unfortunately, because of the unfavourable circumstances of the ring-plane passages in 2025, the next good opportunities will not occur until 2038–9.

In a small telescope, the rings disappear completely for weeks. In a large instrument, however, the ring phenomena are fascinating to watch from night to night. The following excerpts from the author's observing logbook record visual impressions with the 91-cm refractor of the Lick Observatory during the ring-plane passage of 10 August 1995. (In addition to the author, the observers included Stephen James O'Meara, Thomas A. Dobbins and David Graham):

> 10 August. With the Sun about 1½° north of the ring-plane, the rings are shadowed in darkness. Although the seeing is good during late twilight, it deteriorates as the evening progresses, with high winds buffeting the telescope. Still, a ring segment is suspected where the western ansa of the rings should be, with Mimas hanging on it like a tiny jewel.
>
> Later, a do-si-do of moons takes place only 3 arcseconds from the western limb. Tethys is just emerging from a transit across Saturn's face as Mimas, followed by Enceladus, approaches from the opposite direction.
>
> 9 August. In near-perfect seeing, the dark side of Saturn's rings is feebly visible while using averted vision and an occulting bar. The rings are ghostly and nebulous, and two faint starlike beads appear to hang upon the thin thread between Mimas and Saturn.

10 August, about ten hours before the ring-plane crossing. Graham writes, 'I cannot detect any trace of the ring whatsoever other than the shadow of the ring on the globe which is patently obvious.'

11 August, about eleven hours after ring-plane passage, so that the sunlit face of the ring is again in view. The ring is evident as a very thin, clear line – the finest line imaginable – extending with increasing faintness to about two Saturn radii, and of a deep, coppery colour. Graham writes with assurance, 'The ring has returned.' To complete the spectacle, faint satellites lie off the extremity of either ansa. During three hours of observing, the rings continue to brighten perceptibly, until they are even visible in the telescope's 15-cm finderscope.

12–13 August. The rings continue to brighten and widen. They now extend symmetrically about two Saturn radii on either side, tapered like the cross-section of a biconvex lens. The line of sight appears slightly north of the rings' shadow line across the ball. The coppery tone of last night has vanished, and now the rings appear a steady bright yellow.

To the Telescope!

Suffice it to say, there is still useful work to be done on Saturn by the dedicated amateur, especially with regard to time-dependent phenomena such as atmospheric phenomena (spots) and ring spokes. This is even more true now that *Cassini* is no longer monitoring the planet from orbit round Saturn. Though even a casual view is awe-inspiring, the system of Saturn – its globe, rings and numerous satellites – does not yield its secrets easily, and repays close and sustained attention. Enjoy!

APPENDIX I:

SATURN DATA

Orbital Characteristics

Epoch J2000.0

Aphelion 1,514.50 million km (10.1238 AU)

Perihelion 1,352.55 million km (9.0412 AU)

Semi-major axis 1,433.53 million km (9.5826 AU)

Eccentricity 0.0565

Orbital period 29.4571 yr

10,759.22 d

24,491.07 Saturnian solar days

Synodic period 378.09 days

Average orbital speed 9.68 km/s (6.01 m/s)

Mean anomaly 317.020°

Inclination 2.485° to ecliptic

5.51° to Sun's equator

0.93° to invariable plane

Longitude of ascending node 113.665°

Argument of perihelion 339.392°

Physical Characteristics

Mean radius 58,232 km (36,184 mi.)

Equatorial radius 60,268 km (37,449 mi.)

9.449 Earths

Polar radius 54,364 km (33,780 mi.)

8.552 Earths

Flatness 0.09796

Surface area 4.27×10^{10} km^2 (1.65×10^{10} sq mi.)

83.703 Earths

Volume 8.2713×10^{14} km^3 (1.9844×10^{14} cu mi.)

763.59 Earths

Mass 5.6834×10^{26} kg (1.2530×10^{27} lb)

95.159 Earths

Mean density 0.687 g/cm^3 (0.0248 lb/cu in.) (less than water)

Surface gravity 10.44 m/s^2 (34.3 ft/s^2)

1.065 g

Moment of inertia factor 0.210 I/MR^2 estimate

Escape velocity 35.5 km/s (22.1 mi./s)

Sidereal rotation period 10.55 hours (10 hr 33 min)

Equatorial rotation velocity 9.87 km/s (6.13 mi./s;

35,500 km/h)

Axial tilt 26.73° (to orbit)

North pole right ascension 40.589°; 2^h 42^m 21^s

North pole declination 83.537°

Albedo 0.342 (Bond)

0.499 (geometric)

Surface temp. min mean max

1 bar 134 K (−139°C)

0.1 bar 84 K (−189°C)

Apparent magnitude +1.47 to −0.24

Angular diameter 14.5″ to 20.1″ (excludes rings)

Atmosphere:
Surface pressure 140 kPa
Scale height 59.5 km (37.0 mi.)

Composition by volume:
96.3%±2.4% hydrogen (H_2)
3.25%±2.4% helium (He)
0.45%±0.2% methane (CH_4)
0.0125%±0.0075% ammonia (NH_3)
0.0110%±0.0058% hydrogen deuteride (HD)
0.0007%±0.00015% ethane (C_2H_6)

Ices:
ammonia (NH_3)
water (H_2O)
ammonium hydrosulphide (NH_4SH)

Appendix II: Saturn Ring Data

Diameters of the Rings				
Ring	Saturn radii	Kilometres	Miles	Earth diameters
D ring	2.472	148,983	92,573	11.679
C ring	3.054	184,059	114,369	14.429
B ring	3.902	235,166	146,125	18.435
A ring	4.538	273,496	169,943	21.440
F ring	4.652	280,367	174,212	21.979
G ring	5.80	349,554	217,203	27.403
E ring	16	964,288	599,181	75.593

Widths of the Rings				
Ring	Saturn radii	Kilometres	Miles	Earth diameters
D ring	0.126	7,594	4,719	0.595
C ring	0.288	17,357	10,785	1.361
B ring	0.424	25,554	15,878	2.003
A ring	0.242	14,585	9,063	1.143
F ring	–	–	–	–
G ring	0.08	4,821	2,996	0.378
E ring	5	301,340	187,244	23.623

APPENDIX III:
SATURN MOON DATA

Saturn has, as of the time of writing, 62 confirmed satellites (and one more suspected); 53 have received names. Data is given here for the sixteen satellites; the seven that orbit close to the rings (in the cases of Pan and Daphnis, within the A ring); and the nine largest satellites.

Satellite numeral/name	Diameter (km)	Density (gm/cm^3) (water = 1.0 gm/cm^3; ice = 0.931 gm/cm^3)	Visual magnitude	Albedo
I Mimas	396	1.15	12.8	0.6
II Enceladus	504	1.61	11.8	1.0
III Tethys	1066	0.97	10.3	0.8
IV Dione	1123	1.48	10.4	0.6
V Rhea	1529	1.23	9.7	0.6
VI Titan	5149	1.88	9.4	0.2
VII Hyperion	270	0.54	14.4	0.3
VIII Iapetus	1471	1.08	11.0	0.05–0.5
IX Phoebe	213	1.63	16.5	0.08
XVII Pan	28.2 ± 2.6. (34x31x20)	0.42	–	0.5
XXXV Daphnis	7.6 ± 1.6 (9x8x6)	0.34	–	–
XV Atlas	30.2 ± 1.(41x35x19	0.50	–	0.8
XVI Prometheus	86.2 ± 5.4 (130x 114x 106)	0.48	–	0.5
XVII Pandora	81.4 ± 3.6 (130x 114x 106)	0.49	–	0.7
XI Epimetheus	116.2 ± 3.6 (204x186x152)	0.64	–	0.8
X Janus	179.0 ± 2.8	0.54	–	0.9

	Mean distance from planet (10^3 km)	Orbital period (days)	Eccentricity	Orbital inclination (°)	Discovery
	185.5	0.942	0.019 6	1.57	William Herschel, 1789
	238.0	1.370	0.00 0	0.00	William Herschel, 1789
	294.7	1.888	0.00 1	1.09	G. D. Cassini, 1684
	377	2.737	0.002 2	0.03	G. D. Cassini, 1684
	527.1	4.518	0.000 2	0.33	G. D. Cassini, 1672
	1221.8	15.945	0.028 8	0.31	C. Huygens, 1655
	1500.9	21.277	0.023 2	0.62	W. C. Bond, G. P. Bond, W. Lassell, 1848
	3560.9	79.330	0.029 3	8.30	G. D. Cassini, 1671
	12944.3	548	0.1644	178	W. H. Pickering, 1898
	133.584	0.575	0.000035	0.001	M. Showalter, 1990
	136.50	0.594	0.000	0.00	*Cassini*,2006
	137.670	0.602	0.0012	0.003	*Voyager* 1, 1980
	159.5	0.613	0.0022	0.008	*Voyager* 1, 1980
	137.1	0.628	0.0042	0.05	*Voyager* 1, 1980
	151.422	0.6942	0.009	0.34	J. Fountain, S. Larson, 1978
	151.472	0.6945	0.007	0.14	A. Dollfus, 1966

References

Preface

1 Richard A. Proctor, *Saturn and Its System: Containing Discussions of the Motions and Telescopic Appearance of the Planet Saturn, Its Satellites and Rings; the Nature of the Rings; and the Habitability of the Planet*, 2nd edn (London, 1882), p. v.

1 A Pale Yellow Star

1 A.F.O'D. Alexander, *The Planet Saturn: A History of Observation, Theory, and Discovery* (London, 1962), p. 44.
2 A. Pannekoek, *A History of Astronomy* (London, 1961), p. 48.
3 Arthur Koestler, *The Sleepwalkers* (New York, 1959), p. 69.
4 Owen Gingerich, *The Eye of Heaven* (New York, 1993), p. 55.
5 Claudius Ptolemy, *Tetrabiblos*, trans. J. M. Ashland (London, 1822), p. 114.
6 Camille Flammarion, *Popular Astronomy*, trans. J. E. Gore (New York, 1906), p. 432.
7 Ibid., p. 432.
8 Quoted in J.L.E. Dreyer, *Tycho Brahe: A Picture of Scientific Life and Work in the Sixteenth Century* (Edinburgh, 1890), p. 14.
9 Alexander, *The Planet Saturn*, p. 76.
10 Dreyer, *Tycho Brahe*, p. 27.
11 Alexander, *The Planet Saturn*, pp. 76–8.
12 One of the figures known as the conic sections, of which the others are the parabola and the hyperbola, an ellipse is, by definition, the curve generated by a point moving in such a way that the the sum of the distances to two fixed points F1 and F2 (known as the foci of the ellipse) is constant. It follows that an ellipse can be constructed by fixing the ends of a string to points F1 and F2 and moving a pencil at the end of the loop thus formed. The maximum length along the ellipse is known as the major axis, the minimum length the minor axis. If the two ends of the major axis are called A and B, then the eccentricity

e of the ellipse is defined as the ratio of the distance F1C to CA which is also equal to F1F2/AB. The eccentricity ranges from 0, the case of a circle, to 1, a straight line.

2 A Strange Ringed World

1 Galileo Galilei, *Le opere di Galileo Galilei*, ed. Antonio Favaro (Firenza, 1890–1909), vol. X, p. 410.
2 Quoted in A.F.O'D. Alexander, *The Planet Saturn: A History of Observation, Theory and Discovery* (London, 1962), p. 85.
3 Stillman Drake, *Discoveries and Opinions of Galileo* (Garden City, NY, 1957), pp. 143–4.
4 Edward R. Tufte, *Envisioning Information* (Cheshire, CT, 1990), p. 67.
5 T. W. Webb, *Celestial Objects for Common Telescopes*, 6th edn (London, 1917), vol. I, p. 210.
6 J. D. Cassini, *Mémoires de l'Académie Royale des Sciences de Paris* (Paris, 1715), p. 48.
7 P. de Laplace, *Histoire de l'Académie Royale des Sciences de Paris* (Paris, 1787), p. 249.
8 Alexander, *The Planet Saturn*, p. 123.
9 W. Herschel, 'On the Ring of Saturn, and the Rotation of the Fifth Satellite upon Its Axis', *Philosophical Transactions of the Royal Society of London*, vol. LXXII (London, 1792), pp. 1–22. In *Collected Scientific Papers of Sir William Herschel*, ed. J.L.E. Dreyer, vol. I (London, 1912), pp. 429–30.
10 Quoted in F. A. Mitchel, *Ormsby Macknight Mitchel: Astronomer and General* (Boston, MA, and New York, 1887), pp. 130–31.
11 Quoted in Joseph Ashbrook, *The Astronomical Scrapbook: Skywatchers, Pioneers, and Seekers in Astronomy* (Cambridge, MA, 1984), pp. 363–4.
12 G. P. Bond, 'On the Rings of Saturn', *American Journal of Science*, II/12 (1851), pp. 97–105.
13 Benjamin Peirce, 'On the Constitution of Saturn's Ring', *American Journal of Science*, II/12 (1851), pp. 106–8.
14 Neither Coolidge nor Tuttle are well known today. After graduating from Harvard, Tuttle went on to a successful career in the legal profession. On Charles Tuttle and his colourful astronomer brother Horace P. Tuttle, see Richard E. Schmidt, 'The Tuttles of Harvard College Observatory, 1850–1862', *The Antiquarian Astronomer*, VI (2012), pp. 74–104. Coolidge, a grandson of President Thomas Jefferson, joined the Union Army during the Civil War, and was killed in action at the battle of Chickamauga, September 1863. For an appreciation, see W. Sheehan and S. J. O'Meara, 'Phillip Sidney Coolidge, Harvard's Romantic Explorer of the Skies', *Sky and Telescope* (April 1998), pp. 71–5.
15 *Annals of the Astronomical Observatory of Harvard College*, vol. II (Cambridge, MA, 1856–7), p. 50n.

16 Alexander, *The Planet Saturn*, p. 169.
17 Ibid.
18 James Challis to William Thomson, 14 March 1855, in Stephen G. Brush, C.W.F. Everitt and Elizabeth Garber, *Maxwell on Saturn's Ring* (Cambridge, MA, and London, 1983), p. 8.
19 Basil Mahon, *The Man Who Changed Everything: The Life of James Clerk Maxwell* (Chichester, 2003), p. 3.
20 J. C. Maxwell to Lewis Campbell, 28 August 1857, in *The Life of James Clerk Maxwell*, ed. Lewis Campbell and William Garnett (London, 1882), p. 278.
21 J. Clerk Maxwell, *On the Stability of the Motion of Saturn's Rings: An Essay, which obtained the Adams Prize for the Year 1856, in the University of Cambridge* (London, 1859).
22 Daniel Kirkwood, 'On the Nebular Hypothesis, and the Approximate Commensurability of the Planetary Periods', *Monthly Notices of the Royal Astronomical Society*, XXIX (1869), pp. 96–102.
23 James E. Keeler, 'First Observations of Saturn with the 36-inch Equatorial of the Lick Observatory', *Sidereal Messenger*, VII (1888), pp. 79–83.
24 Ibid.
25 W. F. Denning, *Telescopic Work for Starlight Evenings* (London, 1891), p. 195.
26 Richard A. Proctor, *Other Worlds than Ours: The Plurality of Worlds Studied in the Light of Recent Science*, 3rd edn (London, 1872), p. 142.
27 Albert Van Helden, '"Annulo Cingitur": The Solution to the Problem of Saturn', *Journal for the History of Astronomy*, V/3 (1 October 1974), p. 170.
28 Thomas A. Dobbins, Donald C. Parker and Charles F. Capen, *Observing and Photographing the Solar System: A Practical Guide* (Richmond, VA, 1992), p. 107.
29 At the time, telescopes were usually described in terms of their focal lengths rather than their apertures. Herschel's 20-ft (6.1-m) focal length telescope boasted a 47.5-cm diameter mirror, and the 40-ft (12.2-m), a 123-cm mirror. Unfortunately, the seeing conditions in England were rarely good enough to allow the latter to be used to advantage, and most of his routine work was carried out with the former.
30 Alexander, *The Planet Saturn*, p. 128.
31 W. Sheehan, *The Immortal Fire Within: The Life and Work of Edward Emerson Barnard* (Cambridge, 1995), pp. 345–8.

3 Saturn in Depth

1 Richard A. Proctor, *Other Worlds than Ours: The Plurality of Worlds Studied in Light of Recent Science* (New York, 1896), p. 142.
2 W. Sheehan, 'Observations of Saturn in 1992 and 1993 at Pic du Midi and Yerkes', *Journal of the British Astronomical Association*, CIV/4 (1994), pp. 194–6.

3 Carolyn Porco, '*Cassini* at Saturn', *Scientific American*, CCCXVII/4 (October 2017), p. 85.
4 E. E. Barnard, 'White Spot on Saturn', *Astrophysical Journal*, XXIII (1903), pp. 143–4.
5 'Comedian's Big Discovery on Saturn', *Daily Mirror* (8 August 1933).
6 Cheng Li and Andrew P. Ingersoll, 'Moist Convection of Hydrogen Atmospheres and the Frequency of Saturn's Storms', *Nature Geoscience*, VIII/5 (May 2015), pp. 398–402.
7 William B. Hubbard, Michele K. Dougherty, Daniel Gauthier and Robert Jacobson, 'The Interior of Saturn', in *Saturn from Cassini–Huygens*, ed. M. K. Dougherty, Larry W. Esposito and Stamatios M. Krimigis (Dordrecht, 2009), pp. 75–81.
8 A.F.O'D. Alexander, *The Planet Saturn: A History of Observation, Theory and Discovery* (New York, 1962), pp. 371–2.
9 P. M. Celliers, M. Millot, S. Brygoo et al., 'Insulator-metal Transition in Dense Fluid Deuterium', *Science* CCCLXI/6403 (17 August 2018), pp. 677–82.
10 S. J. Bolton, A. Adriani, V. Adumitroaie et al., 'Jupiter's Interior and Deep Atmosphere: The Initial Pole-to-pole Passes with the Juno Spacecraft', *Science*, CCCLVI/6340 (26 May 2017), pp. 821–5.
11 For details, see William Sheehan and Thomas Hockey, *Jupiter* (London, 2018), pp. 153–60.
12 George Biddell Airy, 'On the Mass of Jupiter', *Monthly Notices of the Royal Astronomical Society*, II/20 (12 April 1833), p. 171.
13 See Thomas Gilovich, *How We Know What Isn't So: The Fallibility of Human Reason in Everyday Life* (New York, 1993), p. 29.
14 Frank Wilczek, *Fantastic Realities: 49 Mind Journeys and a Trip to Stockholm* (Hackensack, NJ, 2006), p. 31.
15 A regularly updated list of *Kepler* discoveries is found in the Wikipedia article titled 'List of Exoplanets Discovered Using the *Kepler* Spacecraft', https://en.wikipedia.org.
16 As of April 2018, there were 627 of these systems. A regularly updated list appears in the Wikipedia article titled 'List of Multiplanetary Systems', https://en.wikipedia.org.
17 C. Espillait et al., 'On the Diversity of the Taurus Transitional Disks: UX Tauri A and LkCa 15', *Astrophysical Journal*, DCLXX (2007), pp. L135–8.
18 Konstantin Batygin and Gregory Laughlin, 'Jupiter's Decisive Role in the Inner Solar System's Early Evolution', *Proceedings of the National Academy of Sciences USA*, CXII (2015), pp. 4214–17.
19 Cited in Stephen Brush, *Nebulous Earth: The Origin of the Solar System and the Core of the Earth from Laplace to Jeffreys* (Cambridge, 1996), p. 93.

4 Aesthetic Rubble (the Rings)

1 P. Lowell, 'Memoir on Saturn's Rings', *Memoirs of the Lowell Observatory*, I/2 (1915), p. 3.
2 P. Lowell, 'The Genesis of the Planets', *Journal of the Royal Astronomical Society of Canada*, X/6 (July–August 1916), p. 290.
3 Unfortunately, Lyot did not live to publish his full results, but his colleague Audouin Dollfus did so in *L'Astronomie*, LXVII (1953), p. 3.
4 G. P. Kuiper, 'Report of Commission 16 (Commission pour les Observations Physiques des Planètes et des Satellites)', *Transactions* I.A.U., IX (1955), p. 255.
5 See Audouin Dollfus, 'Saturn's Rings: Divisions, Gaps and Ringlets from Ground-Based Telescopes', in *Anneaux de Planètes: Planetary Rings*, Colloque UAI Toulouse (Paris, 1984). It is also possible that the dazzling brilliance of the image provided by a 5.1-m aperture played a role, since, as the pioneering physicist and perceptual psychologist Gustav Fechner (1801–1887) demonstrated, the eye is capable of distinguishing small gradations in surface brightness far more readily in a comparatively dim image than in an extremely bright one. The late planetary observer and telescope-maker Thomas R. Cave, Jr (1923–2003) used the 500-cm reflector on the planets, and always found its performance quite disappointing compared to the 150-cm and 250-cm reflectors on Mt Wilson.
6 W. C. Livingston, 'Saturn's Rings and Perfect Seeing', *Sky and Telescope* (July 1975), p. 28.
7 A. F. O'D. Alexander, *The Planet Saturn: A History of Observation, Theory and Discovery* (New York, 1962), p. 340.
8 John E. Westfall and William Sheehan, *Celestial Shadows: Eclipses, Transits and Occultations* (New York, 2015), p. 543.
9 Carolyn Porco to W. Sheehan; personal communication, 4 July 2018.
10 Cited in James Elliot and Richard Kerr, *Rings: Discoveries from Galileo to Voyager* (Cambridge, MA, 1984), p. 45.
11 C. C. Lin and F. H. Shu, 'On the Spiral Structure of Disk Galaxies', *Astrophysical Journal*, CXL (1964), pp. 646–55. The waves are also known as quasi-static density waves, or heavy sound waves.
12 Peter Goldreich and Scott Tremaine, 'The Velocity Dispersion in Saturn's Rings', *Icarus*, XXXIV/2 (May 1978), pp. 227–39; 'The Formation of the Cassini Division in Saturn's Rings', *Icarus*, XXXIV/2 (May 1978), pp. 240–53.
13 Quoted in Elliot and Kerr, *Rings*, p. 46.
14 James L. Elliot, Edward W. Dunham and Jessica Mink, 'The Rings of Uranus', *Nature*, CCLXVII/5609 (26 May 1977), pp. 328–30.
15 Scott Tremaine, *Astrophysical Wonders: A Conversation with Scott Tremaine* (Toronto, 2015), p. 31.

16 S. J. O'Meara to W. Sheehan, personal communication, 4 June 2018.
17 Mark Washburn, *Distant Encounters: The Exploration of Jupiter and Saturn* (San Diego, CA, 1983), pp. 199–200.
18 W. Sheehan, July 2016 conversation with Brad Smith. It is pleasant to record that Smith and O'Meara remained on amiable terms, and after the Saturn encounter, Smith called O'Meara at *Sky and Telescope* and challenged him to determine visually the rotation period of Uranus' cloud tops, which was not then accurately known, prior to *Voyager* 2's flyby in January 1986. In this conversation, Smith confided his own plans to perform CCD imaging of Uranus at Cerro Tololo Observatory in Chile, and mentioned that another group of astronomers was planning to do the same at McDonald Observatory in Texas. O'Meara did take up the challenge, and kept up observations with the 23-cm refractor at Harvard, at first without success. Eventually, however, his persistence paid off, and he detected several rogue bright spots on the planet, which allowed him to calculate an average rotation period of 16 hours and 24 minutes. This value – though 'disturbingly discordant' with the 24-hour rotation period inferred by Smith and the McDonald team – proved in the end to be almost spot-on. When *Voyager* 2 arrived at Uranus, it confirmed O'Meara's value to an accuracy of 10 per cent.
19 As noted in David H. Levy, *Clyde Tombaugh: Discoverer of Planet Pluto* (Tucson, AZ, 1991), p. 176.
20 Elliot and Kerr, *Rings*, p. 135.
21 Jeffrey N. Cuzzi, 'Ringed Planets: Still Mysterious – I', *Sky and Telescope* (December 1984), p. 515.

5 *Cassini's* Epic Mission

1 Matthew M. Hedman, Joseph A. Burns, Mark R. Showalter et al., 'Saturn's Dynamic D Ring', *Icarus*, CXCVIII/1 (2007), pp. 89–107.
2 J. N. Cuzzi, A. D. Whizin, R. C. Hogan et al., 'Saturn's F Ring Core: Calm in the Midst of Chaos', American Astronomical Society, Division of Planetary Sciences meeting 44 (October 2012).
3 Robin M. Canup, 'Origin of Saturn's Rings and Inner Moons by Mass Removal from a Lost Titan-sized Satellite', *Nature*, CCCCLXVIII (16 December 2010), pp. 943–6.
4 'A Dozen New Moons of Jupiter Discovered Including One Oddball', Carnegie Institution for Science Press Release, 16 July 2018.
5 Carolyn Porco, 'Cassini at Saturn', *Scientific American*, CCCXVII/4 (October 2017), p. 84.
6 W. W. Morgan, 21 December 1956 entry in personal notebook, Yerkes Observatory archives, University of Chicago.

6 Saturn by Moonlight

1 E. E. Barnard, 'Observations of the Eclipse of Iapetus in the Shadows of the Globe, Crape Ring, and Bright Ring of Saturn, 1889 November 1', *Monthly Notices of the Royal Astronomical Society*, XL (1890), pp. 107–10.
2 Carolyn Porco, 'Cassini at Saturn', *Scientific American*, CCCXVII/4 (October 2017), p. 84.
3 Bonnie Buratti quoted in Charles Q. Choi, 'Weirdly Colored Saturn Moons Linked to Ring Features, NASA's Cassini Revealed', available at www.space.com, accessed 2 April 2019.

7 Observing Saturn

1 John Westfall and William Sheehan, *Celestial Shadows: Eclipses, Transits, and Occultations* (New York and Heidelberg, 2015), p. 224.

Bibliography

Alexander, A.F.O'D., *The Planet Saturn: A History of Observation, Theory and Discovery* [1962] (New York, 1980)
Beatty, J. Kelly, Carolyn Collins Petersen and Andrew Chaikin, eds, *The New Solar System*, 4th edn (Cambridge, MA, 1999)
Brown, Robert H., Jean-Pierre Lebreton and J. Hunter Waite, eds, *Titan from Cassini–Huygens* (New York, 2009)
Brush, Stephen G., C.W.F. Everitt and Elizabeth Garber, eds, *Maxwell on Saturn's Rings* (Cambridge, MA, 1983)
Burns, Joseph A., and Mildred S. Matthews, eds, *Satellites* (Tucson, AZ, 1986)
Burrati, Binnie, et al., 'Close Cassini flybys of Saturn's Ring Moons Pan, Daphnis, Atlas, Pandora, and Epimetheus', *Science*, 28 March 2019
Dobbins, Thomas A., Donald C. Parker and Charles F. Capen, *Observing and Photographing the Solar System: A Practical Guide for the Amateur Astronomer* (Richmond, VA, 1988)
Dougherty, Michele K., Larry W. Esposito and Stamatios M. Krimigis, eds, *Saturn from Cassini–Huygens* (New York, 2009)
Elliot, James, and Richard Kerr, *Rings: Discoveries from Galileo to Voyager* (Cambridge, MA, 1984)
Gehrels, Tom, and Mildred S. Matthews, eds, *Saturn* (Tucson, AZ, 1984)
Greenberg, Richard, and André Brahic, eds, *Planetary Rings* (Tucson, AZ, 1984)
Lorenz, Ralph, and Jacqueline Mitton, *Lifting Titan's Veil: Exploring the Giant Moon of Saturn* (Cambridge, 2002)
Lovett, L., J. Horvath, and J. Cuzzi, *Saturn: A New View* (New York, 2006)
Morrison, David, *Voyages to Saturn*, NASA SP-451 (Washington, DC, 1982)
National Aeronautics and Space Administration, *The Saturn System: Through the Eyes of Cassini* (Washington, DC, 2017)
Schenk, Paul M., Roger N. Clark, Carly J. A. Howett, Anne J. Verbiscer and J. Hunter Waite, eds, *Enceladus and the Icy Moons of Saturn* (Tucson, AZ, 2018)
Sheehan, William, and Thomas Hockey, *Jupiter* (London, 2018)

Spohn, Tilman, Doris Breuer and Torrence Johnson, eds, *Encyclopedia of the Solar System* (Cambridge, MA, 2014)
Westfall, John, and William Sheehan, *Celestial Shadows: Eclipses, Transits, and Occultations* (New York, 2015)

Acknowledgements

Richard Baum, Dale P. Cruikshank, Jeffrey N. Cuzzi, Thomas A. Dobbins, Audouin Dollfus, David Graham, Peter Hingley, Stephen James O'Meara, Donald E. Osterbrock, Carolyn C. Porco, *Cassini* Imaging Team Leader, and Bradford A. Smith, *Voyager* Imaging Team leader, and John Westfall have gone to great trouble over the years to obtain information which would not otherwise have been accessible. Julian Baum has kindly provided two of his beautiful paintings. Michael Conley, to whom the book is dedicated, has been a close friend and collaborator since we were in high school together, and deserves great credit for being ever at the ready both with inspiration and technical know-how.

I owe a great deal to the vision of the editor and founder of the Kosmos series, Peter Morris, to Harry Gilonis, who with patience and good humour helped work through some vexing problems with the illustrations, and publisher Michael Leaman of Reaktion Books. Without their help the book would never have been. Any shortcomings that remain are, needless to say, entirely the fault of the author.

Photo Acknowledgements

The author and publishers wish to express their thanks to the below sources of illustrative material and / or permission to reproduce it. Some locations are also given in the captions for the sake of brevity.

Allegheny Observatory, Pittsburgh: p. 56; collection of the author: pp. 26, 27, 31, 34 (top), 51, 56, 65; courtesy the author: p. 30; from E. E. Barnard, 'Additional Observations of the Disappearances and Reappearances of the Rings of Saturn in 1907–8, made with the 40-inch Refractor of the Yerkes Observatory', *Monthly Notices of the Royal Astronomical Society*, LXVIII/5 (March 1908): p. 68; images by Julian Baum (© Julian Baum): pp. 14, 22; from William Cranch Bond, 'Description of the Observatory at Cambridge', in *Memoirs of the American Academy of Arts and Sciences*, new series, vol. IV, part I (Cambridge, MA, and Boston, 1849): p. 42; from William Cranch Bond, 'Observations on Saturn 1847–1856', in *Annals of the Astronomical Observatory of Harvard College*, vol. II (Cambridge, MA, 1857): p. 48; photo Klaus Brasch: p. 10; from George F. Chambers, *Astronomy* (London, 1910): p. 104; Dale P. Cruikshank: p. 138; from W. R. Dawes, 'On the Ring of Saturn', *Monthly Notices of the Royal Astronomical Society*, XI/2 (December 1851): p. 45; from W. F. Denning, 'The Rev. William Rutter Dawes', *The Observatory*, XXXVI (November 1913): p. 44; from J.L.E. Dreyer's introduction to the *Scientific Papers of William Herschel*, vol. I (London, 1912): p. 66 (upper); ESO, ALMA (ESO/NAOJ/NRAO): p. 101; from Camille Flammarion, *Les Terres et Ciel* (Paris, 1884): p. 21; David Graham/the Saturn Section of the British Astronomical Association: p. 188; Haags Historisch Museum, The Hague: p. 33; Harvard College Observatory, Cambridge, MA: pp. 42, 49; from William Herschel, 'Account of the Discovery of a Sixth and Seventh Satellite of the Planet Saturn; with Remarks on the Construction of its Ring, its Atmosphere, its Rotation on an Axis, and its spheroidal Figure', *Philosophical Transactions of the Royal Society of London* [vol. LXXX (1790)] as reproduced in *Scientific Papers of William Herschel*, vol. I (London, 1912): p. 66 (lower); from William Herschel, 'Observations of a Quintuple Belt on the Planet Saturn',

Phil. Trans., vol. LXXXIV (January 1794): p. 47; Peter Hingley, Royal Astronomical Society: p. 46; from Christian Huygens, *Systema Saturnium* (The Hague, 1659): pp. 31, 34 (top); A. Isella, p. 101; from *Journal of the British Astronomical Association*, vol. XLIV (1933): p. 85; Erich Karkoschka, University of Arizona Lunar and Planetary Lab and NASA/ESA: p. 71 (top); from James E. Keeler, 'A Spectroscopic Proof of the Meteoric Constitution of Saturn's Rings', *Astrophysical Journal*, I (May 1895): p. 52; from Johannes Kepler, *Mysterium Cosmographicum* (Tübingen, 1596): pp. 26, 27; from G. P. Kuiper and B. Middlehurst, *Planets and Satellites* (Chicago, IL, 1961): p. 107; from Percival Lowell, *Memoir on Saturn's Rings* (Lynn, MA, 1915): p. 106; Musée National des Châteaux de Versailles et de Trianon: p. 37; Museo del Prado, Madrid: p. 18; NASA: pp. 60, 92, 116, 117, 121 (foot), 123, 128, 140, 141, 142, 143 (top); NASA/Ted Stryk: p. 114; NASA and The Hubble Heritage Team (STSCI/ AURA): p. 34 (foot); NASA, ESA, J. Clarke (Boston University) and Z. Levay (Space Telescope Science Institute: p. 74; NASA, ESA and the Hubble Heritage Team (STScI/ AURA) 167; NASA, ESA, Space Telescope Institute, M. Mutchler (STSCI), A. Simon (GSFC) and the OPAL Team, J. DePasquale (STScI): p. 185; photo NASA/ JPL-Caltech/Space Science Institute: pp. 6, 57, 70, 71 (foot), 73, 81 (foot), 88, 89, 110 (foot), 113, 124, 125, 131, 132, 134, 135, 136, 137, 142 (foot), 145 (foot), 146, 147, 149, 151, 152 (foot), 152–3 (top), 159, 160, 162, 165, 166, 168, 170, 171, 172, 173, 174, 175, 177, 179, 180; NASA/JPL- Caltech/Space Science Institute/G. Ugarkovic: p. 110 (top); NASA/JPL-Caltech/Space Science Institute/University of Arizona: p. 96, 178; NASA/JPL-Caltech/Space Science Institute/University of Arizona/Andrey Pivovarov: pp. 182–3; NASA/JPL -Caltech/Space Science Institute/ Val Klavans: p. 81 (top); NASA/JPL-Caltech/SwRI/ MSSS/Gerald Eichstad/Sean Doran: p. 93; NASA/JPL/Space Science Institute: pp. 145 (top), 169; NASA/JPL/ University of Arizona: p. 144; Stephen James O'Meara: pp. 17, 121 (top), 127; Observatoire de Paris: p. 39; T.E.R. Phillips, *Hutchinson's Splendour of the Heavens* (London, 1923): p. 36; Carolyn Porco: p. 118; from Richard Proctor, *Saturn and Its System . . .* (London, 1882): pp. 41, 82, 83; B. Saxton (NRAO/ AUI/NSF): p. 101; U.S. Naval Observatory: p. 62; John Westfall: p. 77.

Index

Page numbers in ***bold italics*** refer to illustrations